FOCUS ON EVOLUTION EQUATIONS

FOCUS ON EVOLUTION EQUATIONS

GASTON M. N'GUEREKATA
EDITOR

Nova Science Publishers, Inc.
New York

For permission to use material from this book please contact us:
Telephone 631-231-7269; Fax 631-231-8175
Web Site: http://www.novapublishers.com

LIBRARY OF CONGRESS CATALOGING-IN-PUBLICATION DATA

Focus on evolution equations / [edited by] Gaston M. N'Guerekata.
p. cm.
Includes index.
ISBN 978-1-60021-342-7 (hardcover)
1. Evolution equations. I. N'Guerekata, Gaston M., 1953-
QA377.3.F63 2011
515'.353--dc23

2011028608

Published by Nova Science Publishers, Inc. † New York

CONTENTS

PREFACE

This book presents high-quality research on the theory and methods of linear or nonlinear evolution equations as well as their further applications. Equations dealing with the asymptotic behavior of solutions to evolution equations are included. This book also covers degenerate parabolic equations, abstract differential equations, comments on the Schrodinger equation, solutions in banach spaces, periodic and quasi-periodic solutions, concave Lagragian systems and integral equations.

In Chapter 1 we use variational methods to study the nonexistence of positive solutions for the following nonlinear parabolic partial differential equations:

$$\begin{cases} \dfrac{\partial u}{\partial t} = \Delta_{\mathbb{H}^n}(u^m) + V(w)u^m & \text{in} \quad \Omega \times (0,T), \quad 0 < m < 1, \\ u(w,0) = u_0(w) \geq 0 & \text{in} \quad \Omega \\ u(w,t) = 0 & \text{on} \quad \partial\Omega \times (0,T), \end{cases} \tag{1}$$

and

$$\begin{cases} \dfrac{\partial u}{\partial t} = \Delta_p u + V(w)u^{p-1} & \text{in} \quad \Omega \times (0,T), \quad 1 < p < 2, \\ u(z,l,0) = u_0(z,l) \geq 0 & \text{in} \quad \Omega, \\ u(z,l,t) = 0 & \text{on} \quad \partial\Omega \times (0,T). \end{cases} \tag{2}$$

Here Ω is a bounded domain with smooth boundary in the Heisenberg group and $V \in L^1_{loc}(\Omega)$.

In Chapter 2, various origins of linear and nonlinear Schrödinger equations are discussed in connection with diffusion, hydrodynamics, and fractal structure. The treatment is mainly expository, emphasizing the quantum potential, with a few new observations.

Chapter 3 is concerned with the properties of the class of p-almost automorphic functions recently introduced by the author as well as their applications to abstract differential equations of the form

$$u'(t) = Au(t) + f(t) \quad (*),$$

where A is the infinitesimal generator of a c_0-semigroup $(T(t))_{t\geq 0}$ in a Banach space $\mathbb{X}$ and $f:\alpha\ \mathbb{X}$ is p-almost automophic for some $1\leq p<\infty$. Under suitable assumptions the existence of a p-almost automorphic mild solution to $(*)$ is obtained.

In Chapter 4 we study the model proposed in [9], which describe the glucose homeostasy. The first approximation analysis and a Lyapunov function are developed to obtain the phase portrait. Each result has clinical interpretation which is presented in terms of the dynamical equilibrium between the glucose and hormonal levels.

Chapter 5 is concerned with the existence of invariant manifolds of fully nonlinear evolution equations in Banach spaces of the form

$$\frac{du(t)}{dt} = Au(t), \quad t\geq 0,$$

where the operator A is assumed to be quasi m-accretive, proto-differentiable at the stationary point and generate a C_0-semigroup. The assumption on the proto-differentiability of A allows us to linearize the equations via their Yosida approximation. As a result we obtain conditions for the existence of invariant manifolds that extend the existsting ones to a general class of evolution equations with multi-valued operators. We conclude with some example in PFDEs and PDEs.

We consider in Chapter 6 the equations governing the movement of micropolar fluids in the three dimensional case. We establish a Gevrey regularity result like in the Navier-Stokes equations.

In Chapter 7 we introduce a hypothesis: that $f(\xi)\subset\overline{\text{hull}}\ \{\Phi(\xi)\}$ for a suitable set Φ of Lipshitzian s and use a ness result for parametrized sets of fixpoints to show that this is sufficient to ensure the existence of (generalized) s for the $\dot{x} = \mathrm{A}x + f(x)$.

Chapter 8 is concerned with the almost periodicity, periodicity, and quasi periodicity of bounded solutions and a so-called Massera criterion for the existence of periodic solution of neutral differential equation with piecewise constant argument.

In Chapter 9 we study conditions on the asymptotic equivalence of linear evolution equations in Banach spaces. Besides, we discuss the asymptotic relationship between C_0 semigroups and strongly continuous evolutionary process.

In Chapter 10 we present some theorem on the superposition of an *H*-almost periodic (a.p.) function and the reciprocal of trigonometric polynomial of constant sign. Moreover, we prove theorems on the superposition of a continuous function and an *H*-a.p. function, as well an *N*-a.p. function. Finally, we prove a theorem on the superposition of a differentiable function and a Bohr's a.p. function (*B*-a.p.).

In Chapter 11 we study the asymptotic behaviour of solutions of a concave Lagrangian system, in presence of a discount rate, around a constant solution. Such systems are motivated by the macroeconomic theory. First by using an auxiliary function we exhibit invariant subsets and we obtain a result of unstability. Secondly we use the local theory of hyperbolic fixed points and consequences of the global concavity to obtain an existence and uniqueness result on the bounded solution which starts from a fixed state around the constant solution. We prove that these bounded solutions converge toward the constant solution, and moreover, under a sign condition, that these bounded solutions are optimal.

In Chapter 12 we study the notion of Quasisemigroup of bounded linear operators as a generalization of that of strongly continuous semigroup and give some applications in abstract evolution equations.

The maximal regularity and continuation of the solution and its fractional-order derivative of the solution of the Cauchy problem of the non-homogeneous fractional order evolution equation $D^{\alpha}u(t) = Au(t) + f(t), \alpha \in (0,1)$ have been studied in [6]. In Chapter 13 we study the maximal regularity, continuation of the solution and its derivative and some other properties of the solution of the Cauchy problem of the non-homogeneous fractional order evolutionary integral equation $D^{\alpha}u(t) = \int_0^t h(t-s)Au(s)ds + f(t), \ \alpha \in (0,1)$.

In: Focus on Evolution Equations
Editor: Gaston M. N'Guerekata

ISBN: 978-1-60021-342-7

Chapter 1

NONLINEAR DEGENERATE PARABOLIC EQUATIONS ON THE HEISENBERG GROUP*

Jerome A. Goldstein[1,†] ***and Ismail Kombe***[2,‡]
[1]Department of Mathematical Sciences,
University of Memphis, Memphis TN, U.S.A.
[2]Mathematics Department, Dawson-Loeffler Science & Mathematics Bldg,
Oklahoma City University, Oklahoma City, OK, U.S.A.

Abstract

We use variational methods to study the nonexistence of positive solutions for the following nonlinear parabolic partial differential equations :

$$\begin{cases} \frac{\partial u}{\partial t} = \Delta_{\mathbb{H}^n}(u^m) + V(w)u^m & \text{in} \quad \Omega \times (0,T), \quad 0 < m < 1, \\ u(w,0) = u_0(w) \geq 0 & \text{in} \quad \Omega \\ u(w,t) = 0 & \text{on} \quad \partial\Omega \times (0,T), \end{cases} \tag{0.1}$$

and

$$\begin{cases} \frac{\partial u}{\partial t} = \mathbf{\Delta_p} u + V(w)u^{p-1} & \text{in} \quad \Omega \times (0,T), \quad 1 < p < 2, \\ u(z,l,0) = u_0(z,l) \geq 0 & \text{in} \quad \Omega, \\ u(z,l,t) = 0 & \text{on} \quad \partial\Omega \times (0,T). \end{cases} \tag{0.2}$$

Here Ω is a bounded domain with smooth boundary in the Heisenberg group and $V \in L^1_{\text{loc}}(\Omega)$.

1. Introduction

We study the following nonlinear degenerate parabolic equations:

$$\begin{cases} \frac{\partial u}{\partial t} = \Delta_{\mathbb{H}^n}(u^m) + V(w)u^m & \text{in} \quad \Omega \times (0,T), \quad 0 < m < 1, \\ u(w,0) = u_0(w) \geq 0 & \text{in} \quad \Omega \\ u(w,t) = 0 & \text{on} \quad \partial\Omega \times (0,T), \end{cases} \tag{1.3}$$

---

*Communicated by Gaston M. N'Guérékata

†E-mail address: jgoldste@memphis.edu

‡E-mail address: ikombe@okcu.edu

and

$$\begin{cases} \frac{\partial u}{\partial t} = \mathbf{\Delta_p} u + V(w)u^{p-1} & \text{in} \quad \Omega \times (0,T), \quad 1 < p < 2, \\ u(w,0) = u_0(w) \geq 0 & \text{in} \quad \Omega, \\ u(z,l,t) = 0 & \text{on} \quad \partial\Omega \times (0,T). \end{cases} \tag{1.4}$$

Here Ω is a bounded domain with smooth boundary in the Heisenberg group and $V \in L^1_{\text{loc}}(\Omega)$. The purpose of this paper is to investigate nonexistence of positive solutions of the problems (1.1) and (1.2). In our earlier paper [11] we studied problems (1.1) and (1.2) on Euclidean space. This paper is an extension of [11] to the Heisenberg group case.

We first introduce some basic notation on the Heisenberg group that will be used throughout the article, and further details can be found in [9], [16]. We recall that the Heisenberg group, $\mathbb{H}^n$, is the Lie group whose underlying manifold is $\mathbb{C}^n \times \mathbb{R}$, $n \in \mathbb{N}$, and its group structure is given by

$$w \circ w' = (z + z', l + l' + 2Im(z \cdot z')).$$

Here and in the rest of the paper we identify $\mathbb{C}^n$ with $\mathbb{R}^{2n}$ and setting $z = x + iy$, for a point of $\mathbb{H}^n$ we use the equivalent notations $w = (z,l) = (x,y,l) \in \mathbb{R}^n \times \mathbb{R}^n \times \mathbb{R}$ with $z := (z_1, z_2, ..., z_n) = (x_1, y_1, ..., x_n, y_n)$. Here, " $\cdot$ " denotes the usual inner product in $\mathbb{C}^n$. The Lie algebra of $\mathbb{H}^n$ is generated by the left invariant vector fields

$$X_j = \frac{\partial}{\partial x_j} + 2y_j \frac{\partial}{\partial l}, \quad Y_j = \frac{\partial}{\partial y_j} - 2x_j \frac{\partial}{\partial l}, \quad j = 1, ..., n.$$

The Kohn Laplacian on $\mathbb{H}^n$ is the operator

$$\Delta_{\mathbb{H}^n} = \sum_{j=1}^{n} (X_j^2 + Y_j^2)$$

and the subelliptic gradient is the $2n$ dimensional vector field given by

$$\nabla_{\mathbb{H}^n} = (X_1, ..., X_n, Y_1, ..., Y_n).$$

The subelliptic p-Laplacian is defined by

$$\begin{aligned} \mathbf{\Delta}_p u = & \sum_{j=1}^{n} X_j [\sum_{j=1}^{n} (|X_j u|^2 + |Y_j u|^2)]^{\frac{p-2}{2}} X_j u \\ & + \sum_{j=1}^{n} Y_j [\sum_{j=1}^{n} (|X_j u|^2 + |Y_j u|^2)]^{\frac{p-2}{2}} Y_j u. \end{aligned}$$

Briefly

$$\mathbf{\Delta}_p u = \nabla_{\mathbb{H}^n} \cdot (|\nabla_{\mathbb{H}^n} u|^{p-2} \nabla_{\mathbb{H}^n} u).$$

The distance function on $\mathbb{H}^n$ is defined by

$$\rho = d(w) = d(z,l) = (|z|^4 + l^2)^{\frac{1}{4}}$$

and the distance between two points $w_i = (z_i, l_i) \in \mathbb{H}^n, i = 1, 2$ is given by

$$d(w_1, w_2) = d(w_1^{-1} \circ w_2).$$

The open ball of radius R and center a is given by

$$B_{\mathbb{H}^n}(a, R) := \{w \in \mathbb{H}^n : d(a, w) < R\}.$$

An important group of automorphisms on $\mathbb{H}^n$ is given by the Heisenberg dilations

$$\delta_\lambda(w) = (\lambda z, \lambda^2 l), \quad \lambda > 0, \quad w = (z, l) \in \mathbb{H}^n.$$

The Jacobian determinant of δ_λ is λ^Q where

$$Q = 2n + 2$$

is the homogeneous dimension of $\mathbb{H}^n$.

1.1. Motivation and approaches

Note that if $m = 1$ and $p = 2$ then equations (1.1) and (1.2) reduce to the linear degenerate heat equation with a potential. In this context a sharp result has been obtained by Goldstein and Zhang [12]. In fact, they extended the Baras-Goldstein paper [2] to the Heisenberg group $\mathbb{H}^n$. Here is their result.

Theorem 1.1. *(Goldstein-Zhang)[12]*

(i) Suppose $c > C_*(n) = n^2$. Then the problem

$$\begin{cases} u_t(z,l,t) = \Delta_{\mathbb{H}^n} u(z,l,t) + \frac{c|z|^2}{|z|^4+l^2} u(z,l,t), \\ u(z,l,0) = u_0(z,l), \quad (z,l) = (x,y,l) \in \mathbb{H}^n, \quad t \in (0,T], \quad T > 0 \end{cases} \tag{1.5}$$

has no nonnegative solutions except $u \equiv 0$.

(ii) Suppose $c \leq C_*(n)$. Then the above problem has many positive solutions.

The proof relies on the maximum principle, a scaling argument and a Harnack chain argument. Here $C_*(n) = n^2$ is the sharp Hardy's constant on $\mathbb{H}^n$. The following Hardy inequality on $\mathbb{H}^n$ was proved by Garofalo and Lanconelli [9].

Theorem 1.2. *(Garofalo-Lanconelli)[9]. Let $\phi \in C_c^\infty(\mathbb{H}^n)$, then*

$$C_*(n) \int_{\mathbb{H}^n} \frac{|z|^2}{|z|^4 + l^2} |\phi(z,l)|^2 dz dl \leq \int_{\mathbb{H}^n} |\nabla_{\mathbb{H}^n} \phi(z,l)|^2 dz dl.$$

Here $C_*(n) = n^2$ is the optimal constant.

Note that by Hardy's inequality, the infimum of the expression

$$\frac{\int_{\mathbb{H}^n} |\nabla_{\mathbb{H}^n}\phi(z,l)|^2 dzdl - \int_{\mathbb{H}^n} \frac{c|z|^2}{|z|^4+l^2}|\phi(z,l)|^2 dzdl}{\int_{\mathbb{H}^n} |\phi(z,l)|^2 dzdl}$$

is $-\infty$ or 0, according to $c > C_*(n)$ or $c \leq C_*(n)$.

Goldstein and Zhang [13] extended [12] to general positive singular potentials, i.e.,

$$\begin{cases} \frac{\partial u}{\partial t} = \Delta_{\mathbb{H}^n} u + V(w)u & \text{in} \quad \Omega \times (0,T), \\ u(w,0) = u_0(w) \geq 0 & \text{in} \quad \Omega, \\ u(w,t) = 0 & \text{in} \quad \partial\Omega \times (0,T), \end{cases} \tag{1.6}$$

where $0 \leq V \in L^1_{\text{loc}}(\Omega)$. Using the Cabré and Martel technique [5], they discovered that existence and nonexistence of positive solutions is largely determined by the size of the infimum of the spectrum of the symmetric operator

$$-\Delta_{\mathbb{H}^n} - V,$$

which is defined by

$$\sigma_{\text{inf}}(V;\Omega) := \inf_{0 \not\equiv \phi \in C_c^\infty(\Omega)} \frac{\int_\Omega |\nabla_{\mathbb{H}^n}\phi(w)|^2 dw - \int_\Omega V(w)|\phi(w)|^2 dw}{\int_\Omega |\phi(w)|^2 dw}.$$

The existence and nonexistence of positive solutions of linear parabolic problems with positive potentials on $\mathbb{R}^n$ and $\mathbb{H}^n$ is now well understood with these results. Let us mention that we also studied the linear heat equation with a highly singular, oscillatory potential on $\mathbb{R}^n$ [10], [14].

We should mention that there are also significant results for nonlinear parabolic equations with singular potentials on Euclidean space. Garcia and Peral [8] and Aguilar and Peral [1] found an analogue of the Baras-Goldstein paper [2] for the p-Laplacian. Recently, Goldstein and Kombe [11] studied problems (1.1) and (1.2) on a bounded domain with smooth boundary in $\mathbb{R}^n$. In this present paper we extended the results of [11] to the Heisenberg group $\mathbb{H}^n$. We will focus on some critical potential. As a concrete example we will treat the potential of (1.3) and highly singular, oscillatory potentials. We prove that the nonexistence of positive solutions is intimately related with Hardy's inequality on $\mathbb{H}^n$.

Throughout the paper, Ω is a bounded domain in $\mathbb{H}^n$ with a smooth boundary, V is a real valued and locally integrable function on Ω. We are concerned with the general positive local solution of the problem (1.1) [resp. (1.2)].

Definition. By a *positive local solution continuous* off of $\mathcal{K}$, we mean

(i) $\mathcal{K}$ is a closed Lebesgue null subset of Ω,

(ii) $u : [0,T) \longrightarrow L^1(\Omega)$ is continuous for some $T > 0$,

(iii) $(w,t) \longrightarrow u(w,t) \in C((\Omega \setminus \mathcal{K}) \times (0,T))$,

(iv) $u(w,t) > 0$ on $(\Omega \setminus \mathcal{K}) \times (0,T)$,

(v) $\lim_{t\to 0} u(.,t) = u_0$ in the sense of distributions,

(vi) $\nabla_{\mathbb{H}^n} u \in L^2_{\text{loc}}(\Omega)$ [resp. $\nabla_{\mathbb{H}^n} u \in L^p_{loc}(\Omega)$] and u is a solution in the sense of distributions of the PDE.

Remark : If $0 < a < b < T$ and $\mathcal{K}_o$ is a compact subset of $\Omega \setminus \mathcal{K}$, then $u(w,t) \geq \epsilon_1 > 0$ for $(w,t) \in \mathcal{K}_o \times [a,b]$ for some $\epsilon_1 > 0$. We can weaken (iii), (iv) to be

(iii)' $u(w,t)$ is positive and locally bounded on $(\Omega \setminus \mathcal{K}) \times (0,T)$,

(iv)' $\frac{1}{u(w,t)}$ is locally bounded on $(\Omega \setminus \mathcal{K}) \times (0,T)$.

If a solution satisfies (i), (ii), (iii)', (iv)', (v), and (vi) then we call it a " general positive local solution off of $\mathcal{K}$ ". This is more general than a positive local solution continuous off of $\mathcal{K}$. If $\mathcal{K} = \emptyset$, we simply call u " general positive local solution ".

2. The Filtration Equation $(0 < m < 1)$

Equation (1.1) is a perturbation of the filtration equation $\partial_t u = \Delta_{\mathbb{H}^n}(\varphi(u))$ when the monotone function is given by $\varphi(s) = s_+^m$ with $0 < m < 1$; sometimes this is called the fast diffusion equation. Here we will consider the filtration equation with a singular lower order term.

$$\begin{cases} \frac{\partial u}{\partial t} = \Delta_{\mathbb{H}^n}(u^m) + V(w)u^m & \text{in} \quad \Omega \times (0,T), \quad 0 < m < 1, \\ u(w,0) = u_0(w) \geq 0 & \text{in} \quad \Omega, \\ u(w,t) = 0 & \text{on} \quad \partial\Omega \times (0,T). \end{cases} \tag{2.7}$$

The main result of this section is the following theorem.

Theorem 2.3. *Let $n \in \mathbb{N}$, $\frac{n}{n+1} \leq m < 1$ and $V \in L^1_{loc}(\Omega \setminus \mathcal{K})$, where $\mathcal{K}$ is a closed Lebesgue null subset of Ω, where Ω is a smooth bounded domain in $\mathbb{H}^n$. If*

$$\begin{aligned} \sigma_{\inf}((1-\epsilon)V;\Omega) :&= \inf_{0 \not\equiv \phi \in C_c^\infty(\Omega\setminus\mathcal{K})} \frac{\int_\Omega |\nabla_{\mathbb{H}^n}\phi|^2 dw - \int_\Omega (1-\epsilon)V|\phi|^2 dw}{\int_\Omega |\phi|^2 dw} \\ &= -\infty \end{aligned}$$

for some $\epsilon > 0$, then the problem (2.1) has no general positive local solution off of $\mathcal{K}$.

Proof. We argue by contradiction. Given any $T > 0$, let $u : [0,T) \longrightarrow L^1(\Omega)$ be a general positive local solution to (2.1) in $(\Omega \setminus \mathcal{K}) \times (0,T)$ with $u_0 \geq 0$ but not identically zero.

Multiply both sides of (2.1) by the test function ϕ^2/u^m and integrate over Ω, where $\phi \in C_c^\infty(\Omega \setminus \mathcal{K})$; the result is

$$\frac{1}{1-m}\frac{d}{dt}\int_\Omega u^{1-m}\phi^2(w)dw = \int_\Omega (\Delta_{\mathbb{H}^n} u)\frac{\phi^2}{u}dw + \int_\Omega V(w)\phi^2(w)dw. \tag{2.8}$$

Integration by parts gives

$$\begin{aligned}&\frac{1}{1-m}\frac{d}{dt}\int_\Omega u^{1-m}\phi^2(w)dw\\&=\int_\Omega(\frac{\phi^2}{u^2}|\nabla_{\mathbb{H}^n}u|^2-2\frac{\phi}{u}\nabla_{\mathbb{H}^n}u\cdot\nabla_{\mathbb{H}^n}\phi)dw+\int_\Omega V(w)\phi^2(w)dw.\end{aligned}\tag{2.9}$$

A direct computation shows that

$$\int_\Omega(\frac{\phi^2}{u^2}|\nabla_{\mathbb{H}^n}u|^2-2\frac{\phi}{u}\nabla_{\mathbb{H}^n}u\cdot\nabla_{\mathbb{H}^n}\phi)dw\geq-\int_\Omega|\nabla_{\mathbb{H}^n}\phi|^2dw.\tag{2.10}$$

Substituting (2.4) into (2.3), we obtain

$$\int_\Omega V(w)\phi^2(w)dw-\int_\Omega|\nabla_{\mathbb{H}^n}\phi|^2dw\leq\frac{1}{1-m}\frac{d}{dt}\int_\Omega u^{1-m}\phi^2(w)dw.\tag{2.11}$$

Integrating from t_1 to t_2 $(0<t_1<t_2<T)$, we obtain

$$\begin{aligned}&\int_\Omega V(w)\phi^2(w)dw-\int_\Omega|\nabla_{\mathbb{H}^n}\phi|^2dw\\&\leq K_1\int_\Omega(u^{1-m}(w,t_2)-u^{1-m}(w,t_1))\phi^2dw,\end{aligned}\tag{2.12}$$

where

$$K_1=\frac{1}{(1-m)(t_2-t_1)}.$$

Using Jensen's inequality for concave functions yields

$$\int_\Omega\Big(u(w,t_i)\Big)^{\frac{(1-m)(2n+2)}{2}}dw\leq C(|\Omega|)\Big(\int_\Omega u(w,t_i)dw\Big)^{\frac{(1-m)(2n+2)}{2}}<\infty,$$

since we assumed that $\frac{n}{n+1}\leq m$. Here we use the fact that Ω is bounded, whence $|\Omega|$ is finite. Therefore

$$u^{1-m}(w,t_i)\in L^{n+1}(\Omega).$$

By Proposition A.1 in the Appendix, we have

$$\begin{aligned}&K_1\int_\Omega(u^{1-m}(w,t_2)-u^{1-m}(w,t_1))\phi^2dw\\&\leq\frac{\epsilon}{1-\epsilon}\int_\Omega|\nabla_{\mathbb{H}^n}\phi|^2dw+C(\epsilon)\int_\Omega\phi^2dw,\end{aligned}\tag{2.13}$$

for any given $\epsilon>0$. Substituting (2.7) into (2.6) gives

$$\begin{aligned}&\int_\Omega V(w)\phi^2(w)dw-\int_\Omega|\nabla_{\mathbb{H}^n}\phi|^2dw\\&\leq\frac{\epsilon}{1-\epsilon}\int_\Omega|\nabla_{\mathbb{H}^n}\phi|^2dw+C(\epsilon)\int_\Omega\phi^2dw.\end{aligned}\tag{2.14}$$

Therefore

$$\inf_{0\not\equiv\phi\in C_c^\infty(\Omega\backslash\mathcal{K})} \frac{\int_\Omega |\nabla_{\mathbb{H}^n}\phi|^2 dw - \int_\Omega (1-\epsilon)V(w)\phi^2(w)dw}{\int_\Omega \phi^2 dw} \geq -(1-\epsilon)C(\epsilon) \tag{2.15}$$
$$> -\infty.$$

This contradicts our assumption. The proof of the Theorem 2.1 is now complete. □

Singular potentials

We now consider the following potentials for the next corollaries :

$$V(w) = \frac{c|z|^2}{|z|^4 + l^2}, \tag{2.16}$$

and

$$V(w) = \frac{c|z|^2}{|z|^4 + l^2} + \frac{\beta|z|^2}{|z|^4 + l^2}\sin\Big(\frac{1}{(|z|^4 + l^2)^{\frac{\alpha}{4}}}\Big), \tag{2.17}$$

where $c > 0$, $\beta \in \mathbb{R} \setminus \{0\}$ and $\alpha > 0$. Notice that when $|\beta| > c$, V has very large positive and negative parts, in particular, it oscillates wildly, but important cancellations occur between the positive and the negative parts in the quadratic form. Therefore, nonexistence of positive solutions only depends on the size of c. Before proceeding to the corollaries, we want to write some known facts.

Lemma 2.1. *Let $\phi = \phi(\rho)$ be a function of ρ only on $\mathbb{H}^n$, where $\rho = ((x^2+y^2)^2+l^2)^{1/4} = (r^4 + l^2)^{1/4}$. Then*

$$|\nabla_{\mathbb{H}^n}\phi|^2 = \frac{r^2}{\rho^2}|\phi'(\rho)|^2.$$

Proof. For completeness we give the proof. We have

$$|\nabla_{\mathbb{H}^n}\phi|^2 = \sum_{i=1}^n \Big((X_i\phi)^2 + (Y_i\phi)^2\Big),$$

where

$$X_i\phi = (\frac{\partial\rho}{\partial x_i} + 2y_i\frac{\partial\rho}{\partial l})\phi'(\rho), \quad Y_i = (\frac{\partial\rho}{\partial y_i} - 2x_i\frac{\partial\rho}{\partial l})\phi'(\rho).$$

By straightforward computations, we have

$$\begin{aligned}\frac{\partial\rho}{\partial x_i} &= \Big[(x^2+y^2)^2 + l^2\Big]^{-3/4}(x^2+y^2)x_i\phi'(\rho)\\ &= (r^4+l^2)^{-3/4}x_i r^2\phi'(\rho)\\ &= \frac{r^2}{\rho^3}x_i\phi'(\rho),\end{aligned}$$

$$\begin{aligned}\frac{\partial \rho}{\partial y_i} &= \Big[(x^2+y^2)^2+l^2\Big]^{-3/4}(x^2+y^2)y_i\phi'(\rho)\\ &= (r^4+l^2)^{-3/4}y_i r^2\phi'(\rho)\\ &= \frac{r^2}{\rho^3}y_i\phi'(\rho),\end{aligned}$$

and

$$\begin{aligned}\frac{\partial \phi}{\partial l} &= \Big[(x^2+y^2)^2+l^2\Big]^{-3/4}\frac{l}{2}\phi'(\rho)\\ &= \frac{l}{2}\rho^{-3}\phi'(\rho).\end{aligned}$$

Clearly

$$X_i\phi = (\frac{r^2}{\rho^3}x_i + 2y_i\frac{l}{2}\rho^{-3})\phi'(\rho),$$

$$Y_i\phi = (\frac{r^2}{\rho^3}y_i - 2x_i\frac{l}{2}\rho^{-3})\phi'(\rho).$$

Therefore

$$|\nabla_{\mathbb{H}^n}\phi|^2 = \sum_{i=1}^{n}(X_i\phi)^2 + (Y_i\phi)^2 = \frac{r^2}{\rho^2}|\phi'(\rho)|^2.$$

□

Let $D = B_{\mathbb{H}^n}(0,R_2)\setminus\overline{B_{\mathbb{H}^n}(0,R_1)}$ be an annulus with $0 \le R_1 < R_2$, $u \in L^1(D)$. In order to compute $\int_D \phi(w)dw$, we use the following spherical transformation (see [3], [4], [6]).

Let $w = (x,y,l) := \Phi(\rho,\theta,\theta_1,...,\theta_{2n-1})$ and

$$\begin{aligned}x_1 &= \rho(\sin\theta)^{1/2}\cos\theta_1,\\ y_1 &= \rho(\sin\theta)^{1/2}\sin\theta_1\cos\theta_2,\\ &\dots\\ x_{n-1} &= \rho(\sin\theta)^{1/2}\sin\theta_1\sin\theta_2...\cos\theta_{2n-3},\\ y_{n-1} &= \rho(\sin\theta)^{1/2}\sin\theta_1\sin\theta_2...\sin\theta_{2n-3}cos\theta_{2n-2},\\ x_n &= \rho(\sin\theta)^{1/2}\sin\theta_1\sin\theta_2...\sin\theta_{2n-3}\sin\theta_{2n-2}\cos\theta_{2n-1},\\ y_n &= \rho(\sin\theta)^{1/2}\sin\theta_1\sin\theta_2...\sin\theta_{2n-3}\sin\theta_{2n-2}\sin\theta_{2n-1},\\ l &= \rho^2\cos\theta,\end{aligned}$$

for $R_1 < \rho < R_2$, $\theta \in]0,\pi[$, $\theta_i \in]0,\pi[$ for $i = 1,...,2n-2$ and $\theta_{2n-1} \in]0,2\pi[$.

Here

$$\rho = ((x^2+y^2)^2+l^2)^{1/4},$$
$$\theta = \arcsin(\frac{x^2+y^2}{(x^2+y^2)^2+l^2)^{1/2}}),$$
$$\theta_1 = \cos^{-1}\Big(\frac{x_1}{(x^2+y^2)^{1/2}}\Big),$$
$$\ldots$$
$$\theta_{2n-2} = \cos^{-1}\Big(\frac{y_{n-1}}{(x^2+y^2)^{1/2}\prod_{i=1}^{2n-3}\sin\theta_i}\Big),$$
$$\theta_{2n-1} = \cos^{-1}\Big(\frac{x_{2n-2}}{(x^2+y^2)^{1/2}\prod_{i=1}^{2n-2}\sin\theta_i}\Big).$$

We have the volume element

$$dw = dxdydl = \det J(\Phi)d\theta_1...\theta_{2n-2}d\theta d\rho$$

where $J(\Phi)$ is the Jacobian of Φ and

$$\det J(\Phi) = \rho^{2n+1}\sin^{n-1}\theta\sin^{2n-2}\theta_1...\sin\theta_{2n-2}.$$

Let ϕ be a "cylindrical function", i.e., $\phi(w) = \phi(r,l)$ where r is the Euclidean norm of $z = (x,y) \in \mathbb{R}^{2n}$. Then

$$\int_D \phi(r,l)dw = \Gamma_n \int_0^\pi d\theta \int_{R_1}^{R_2} \rho^{2n+1}(\sin\theta)^{n-1}\phi(\rho^2\sin\theta, \rho^2\cos\theta)d\rho,$$

where

$$\Gamma_n := \int_0^\pi d\theta_1 \int_0^\pi d\theta_2 ... \int_0^\pi d\theta_{2n-2}\int_0^{2\pi} d\theta_{2n-1}\sin^{2n-2}\theta_1...\sin\theta_{2n-2}$$

is the $2n$ dimensional Lebesgue surface measure of the unit sphere in $\mathbb{R}^{2n+1}$.

If $\phi(w) = \frac{r^2}{\rho^2}\varphi(\rho) = \sin\theta\varphi(\rho)$, then we have

$$\int_D \phi(w)dw = s_n \int_{R_1}^{R_2} \rho^{2n+1}\varphi(\rho)d\rho \qquad (2.18)$$

where

$$s_n = \Gamma_n \int_0^\pi (\sin\theta)^n d\theta.$$

Corollary 2.1. *Let $0 \in \Omega$, $C_*(n) = n^2$ and $V(w) = \frac{c|z|^2}{|z|^4+l^2}$. Then the problem (2.1) has no general positive local solution off of $\mathcal{K}$ if $c > C_*(n)$ and $\frac{n}{n+1} \le m < 1$.*

Proof. Given $\epsilon > 0$, we define the radial function, $\phi \in C_c^1(\Omega) \bigcap W^{1,\infty}(\Omega)$, by

$$\phi(\rho) = \begin{cases} \epsilon^{-n-1}\rho & \text{if} \quad 0 \le \rho \le \epsilon \\ \rho^{-n} & \text{if} \quad \epsilon \le \rho \le 1 \\ 2-\rho & \text{if} \quad 1 \le \rho \le 2 \\ 0 & \text{if} \quad \rho \ge 2. \end{cases} \qquad (2.19)$$

We are assuming that $0 \in \Omega$. Without loss of generality we assume that $B_{\mathbb{H}^n}(0,2) = \{w \in \mathbb{H}^n : d(0,w) < 2\} \subset \Omega$; if not , we simply redefine ϕ, replacing 2 by R where $B_{\mathbb{H}^n}(0,R) \subset \Omega$. This only results in notational changes in the proof that follows. Then we assume (without loss of generality) that $\phi \in C_c^\infty(\Omega)$.

We want to show that

$$\sigma_{\inf}(V;\Omega) := \inf_{0 \not\equiv \psi \in C_c^\infty(\Omega \setminus \mathcal{K})} \frac{\int_\Omega |\nabla_{\mathbb{H}^n}\psi|^2 dzdl - \int_\Omega \frac{c|z|^2}{|z|^4+l^2}|\psi|^2 dzdl}{\int_\Omega |\psi|^2 dw} = -\infty. \tag{2.20}$$

Let's compute $\sigma_{\inf}(V;\Omega)$.

Using Lemma 2.1, we get, for ϕ as in (2.13),

$$|\nabla_{\mathbb{H}^n}\phi(\rho)|^2 = \begin{cases} \epsilon^{-2n-2}\frac{r^2}{\rho^2} & \text{if } \ 0 \le \rho < \epsilon \\ n^2\rho^{-2n-2}\frac{r^2}{\rho^2} & \text{if } \ \epsilon < \rho < 1 \\ \frac{r^2}{\rho^2} & \text{if } \ 1 < \rho < 2 \\ 0 & \text{if } \ \rho > 2. \end{cases} \tag{2.21}$$

By straightforward computations and using (2.12),

$$\begin{aligned} \int_\Omega |\nabla_{\mathbb{H}^n}\phi|^2 dw &= s_n\Big[\int_0^\epsilon \epsilon^{-2n-2}\rho^{2n+1}d\rho + \int_\epsilon^1 n^2\rho^{-2n-2}\rho^{2n+1}d\rho \\ &\quad + \int_1^2 \rho^{2n+1}d\rho\Big] \\ &= s_n\Big[-n^2\log\epsilon + \frac{2^{2n+2}}{2n+2}\Big]. \end{aligned} \tag{2.22}$$

Since

$$V(z,l) = \frac{|z|^2}{|z|^4+l^2} = \frac{r^2}{\rho^4},$$

we write

$$\begin{aligned} \int_\Omega \frac{|z|^2}{|z|^4+l^2}\phi^2 dw &= s_n\Big[\epsilon^{-2n-2}\int_0^\epsilon \rho^{2n+1}d\rho + \int_\epsilon^1 \frac{1}{\rho}d\rho \\ &\quad + \int_1^2 (2-\rho)^2\rho^{2n-1}d\rho\Big] \\ &= s_n\Big[-\log\epsilon + \frac{2^{2n+2}-2n^2-5n-4}{n(2n+1)(2n+2)}\Big]. \end{aligned} \tag{2.23}$$

Next,

$$\begin{aligned}\int_\Omega \phi^2(w)dw &= \gamma_n\Big[\epsilon^{-2n-2}\int_0^\epsilon \rho^{2n+3}d\rho + \int_\epsilon^1 \rho d\rho \\ &\quad + \int_1^2 (2-\rho)^2\rho^{2n+1}d\rho\Big] \\ &= \gamma_n\Big[\frac{\epsilon^2}{2n+4} + \frac{1}{2} - \frac{\epsilon^2}{2} \\ &\quad + \frac{2^{2n+5} - n2^{2n+5} - 4n^2 - 10n - 12}{(2n+2)(2n+3)(2n+4)}\Big] \\ &= \gamma_n\Big[\frac{1}{2} + K(n)\Big] \quad \text{as} \quad \epsilon \longrightarrow 0.\end{aligned} \tag{2.24}$$

where

$$K(n) = \frac{2^{2n+5} - n2^{2n+5} - 4n^2 - 10n - 12}{(2n+2)(2n+3)(2n+4)}$$

and $\gamma_n = \Gamma_n \int_0^\pi (\sin\theta)^{n-1}d\theta$.

Substituting (2.16), (2.17) and (2.18) into the Rayleigh quotient,

$$\mathcal{R} = \frac{\int_\Omega |\nabla_{\mathbb{H}^n}\phi|^2 dzdl - \int_\Omega \frac{c|z|^2}{|z|^4+l^2}|\phi|^2 dzdl}{\int_\Omega |\phi|^2 dzdl}$$

$$\begin{aligned}&= \frac{s_n\Big[-n^2\log\epsilon + \frac{2^{2n+2}}{2n+2} - c(-\log\epsilon + \frac{2^{2n+2}-2n^2-5n-4}{n(2n+1)(2n+2)})\Big]}{\gamma_n\Big[\frac{1}{2}+K(n)\Big]} \\ &= \frac{s_n\Big[(-n^2+c)\log\epsilon + \frac{2^{2n+2}}{2n+2} - c(\frac{2^{2n+2}-2n^2-5n-4}{n(2n+1)(2n+2)})\Big]}{\gamma_n\Big[\frac{1}{2}+K(n)\Big](1+o(1))}.\end{aligned} \tag{2.25}$$

Therefore taking limits as $\epsilon \longrightarrow 0^+$, we find that the right hand-side of (2.19) can become negative and arbitrarily large in magnitude, i.e.,

$$\sigma_{\inf}(V;\Omega) := \inf_{0\neq\phi\subset(\Omega\setminus\mathcal{K})} \frac{\int_\Omega |\nabla_{\mathbb{H}^n}\phi|^2 dzdl - \int_\Omega \frac{c|z|^2}{|z|^4+l^2}|\phi|^2 dzdl}{\int_\Omega |\phi|^2 dzdl} = -\infty \tag{2.26}$$

The proof of Corollary 2.1 is now complete. □

Corollary 2.2. *Let $0 \in \Omega$, $C_* = n^2$, and $V(z,l)$ be defined by (2.11). Then the problem (2.1) has no general positive local solution off of $\mathcal{K}$ if $c > C_*(n)$ and $\frac{n}{n+1} \leq m < 1$.*

Proof: We define the radial function, $\phi \in C_c^1(\Omega)\bigcap W^{1,\infty}(\Omega)$, by

$$\phi(r) = \begin{cases} \epsilon^{-n-1}\rho & \text{if} \quad 0 \leq \rho \leq \epsilon \\ \rho^{-n} & \text{if} \quad \epsilon \leq \rho \leq 1 \\ 2-\rho & \text{if} \quad 1 \leq \rho \leq 2 \\ 0 & \text{if} \quad \rho \geq 2. \end{cases} \tag{2.27}$$

Without loss of generality we assume that $B_{\mathbb{H}^n}(0,2) = \{w \in \mathbb{H}^n : d(0,w) < 2\} \subset \Omega$.

We want to show that

$$\sigma_{\inf}(V;\Omega) := \inf_{0\not\equiv\psi\in(\Omega\backslash\mathcal{K})} \frac{\int_\Omega |\nabla_{\mathbb{H}^n}\psi|^2 dzdl - \int_\Omega V(z,l)\psi^2 dzdl}{\int_\Omega |\psi|^2 dzdl} = -\infty. \tag{2.28}$$

We already computed most of the relevant integrals, and they are

$$\begin{aligned}\int_\Omega |\nabla_{\mathbb{H}^n}\phi|^2 dw &= s_n\Big[\int_0^\epsilon \epsilon^{-2n-2}\rho^{2n+1}d\rho + \int_\epsilon^1 n^2\rho^{-2n-2}\rho^{2n+1}d\rho \\ &\quad + \int_1^2 \rho^{2n+1}d\rho\Big] \\ &= s_n\Big[-n^2\log\epsilon + \frac{2^{2n+2}}{2n+2}\Big].\end{aligned} \tag{2.29}$$

Next,

$$\begin{aligned}\int_\Omega \frac{|z|^2}{|z|^4+l^2}\phi^2 dw &= s_n\Big[\epsilon^{-2n-2}\int_0^\epsilon \rho^{2n+1}d\rho + \int_\epsilon^1 \frac{1}{\rho}d\rho \\ &\quad + \int_1^2 (2-\rho)^2\rho^{2n-1}d\rho\Big] \\ &= s_n\Big[-\log\epsilon + \frac{2^{2n+2}-2n^2-5n-4}{n(2n+1)(2n+2)}\Big].\end{aligned} \tag{2.30}$$

Next, we have

$$\begin{aligned}\int_\Omega \phi^2(w)dw &= \gamma_n\Big[\epsilon^{-2n-2}\int_0^\epsilon \rho^{2n+3}d\rho + \int_\epsilon^1 \rho d\rho \\ &\quad + \int_1^2 (2-\rho)^2\rho^{2n+1}d\rho\Big] \\ &= \gamma_n\Big[\frac{\epsilon^2}{2n+4} + \frac{1}{2} - \frac{\epsilon^2}{2} \\ &\quad + \frac{2^{2n+5} - n2^{2n+5} - 4n^2 - 10n - 12}{(2n+2)(2n+3)(2n+4)}\Big] \\ &= \gamma_n\Big[\frac{1}{2} + K(n)\Big] \quad \text{as} \quad \epsilon \longrightarrow 0\end{aligned} \tag{2.31}$$

where

$$K(n) = \frac{2^{2n+5} - n2^{2n+5} - 4n^2 - 10n - 12}{(2n+2)(2n+3)(2n+4)}.$$

We only need to compute

$$\int_\Omega \frac{|z|^2}{|z|^4+l^2}\sin\Big(\frac{1}{(|z|^4+l^2)^{\frac{\alpha}{4}}}\Big)\phi^2 dw.$$

Using (2.12) we have

$$I = \int_\Omega \frac{|z|^2}{|z|^4 + l^2} \sin\Big(\frac{1}{(|z|^4 + l^2)^\alpha}\Big)\phi^2 dw = s_n\Big[\int_0^2 \rho^{2n-1} \sin(\frac{1}{\rho^\alpha})\phi^2 d\rho\Big]$$
$$= s_n\Big[\epsilon^{-2n-2}\int_0^\epsilon \rho^{2n+1} \sin(\frac{1}{\rho^\alpha})d\rho + \int_\epsilon^1 \frac{1}{\rho}\sin(\frac{1}{\rho^\alpha})d\rho$$
$$+ \int_1^2 (2-\rho)^2\rho^{2n-1}\sin(\frac{1}{\rho^\alpha})d\rho\Big].$$

Let

$$A = \int_0^\epsilon \rho^{2n+1}\sin(\frac{1}{\rho^\alpha})d\rho,$$
$$B = \int_\epsilon^1 \frac{1}{\rho}\sin(\frac{1}{\rho^\alpha})d\rho,$$
$$C = \int_1^2 (2-\rho)^2 \sin(\frac{1}{\rho^\alpha})d\rho;$$

then $I = s_n(A + B + C)$. Using a change of variable, we write

$$A = \frac{1}{\alpha}\int_{\epsilon^{-\alpha}}^\infty s^{\frac{-(2n+\alpha+2)}{\alpha}} \sin(s)ds.$$

We let $g_1(s) = s^{\frac{-(2n+\alpha+2)}{\alpha}}$; $0 < g_1 \in C[\epsilon^{-\alpha}, \infty)$. It is a positive decreasing function and approaches zero as $s \longrightarrow \infty$. Using the alternating series method (see [11]) we obtain

$$|A| \leq \frac{4}{\alpha}(1 + o(1)) \quad \text{as} \quad \epsilon \longrightarrow 0^+. \tag{2.32}$$

Next,

$$B = \int_\epsilon^1 \frac{1}{\rho}\sin(\frac{1}{\rho^\alpha})d\rho.$$

Using a change of variable, we write

$$B = \frac{1}{\alpha}\int_1^{\epsilon^{-\alpha}} \frac{\sin(s)}{s} ds.$$

Clearly

$$|B| \leq \frac{4}{\alpha}(1 + o(1)) \quad \text{as} \quad \epsilon \longrightarrow 0^+. \tag{2.33}$$

Next,

$$C = \int_1^2 (2-\rho)^2 \sin(\frac{1}{\rho^\alpha})d\rho.$$

As an upper bound we obtain

$$|C| \leq \frac{1}{3}. \tag{2.34}$$

We conclude that

$$|I| \leq s_n \Big[\frac{8}{\alpha}(1 + o(1)) + \frac{1}{3}\Big]. \tag{2.35}$$

Finally, substituting (2.23), (2.24), (2.25) and (2.29) into the Rayleigh quotient,

$$\begin{aligned} \mathcal{R} &= \frac{\int_\Omega (|\nabla_{\mathbb{H}^n}\phi|^2 - (\frac{c|z|^2}{|z|^4+l^2} + \frac{|z|^2}{|z|^4+l^2}\sin(\frac{1}{(|z|^4+l^2)^\alpha})))\phi^2 dw}{\int_\Omega \phi^2 dw} \\ &\leq \frac{s_n\Big[(-n^2+c)\log\epsilon + \frac{2^{2n+2}}{2n+2} - c(\frac{2^{2n+2}-2n^2-5n-4}{n(2n+1)(2n+2)})}{\gamma_n\Big[\frac{1}{2} + K(n)\Big]} \\ &+ \frac{-\beta(\frac{8}{\alpha}(1+o(1)) + \frac{1}{3})\Big]}{\gamma_n\Big[\frac{1}{2} + K(n)\Big]} \end{aligned} \tag{2.36}$$

Therefore taking limits as $\epsilon \longrightarrow 0^+$, we find that the right hand-side of (2.30) can become negative and arbitrarily large in magnitude, i.e.,

$$\sigma_{\text{inf}}(V;\Omega) := \inf_{0 \not\equiv \phi \in (\Omega \backslash \mathcal{K})} \frac{\int_\Omega |\nabla_{\mathbb{H}^n}\phi|^2 dzdl - \int_\Omega V(z,l)\phi^2 dzdl}{\int_\Omega |\phi|^2 dzdl} = -\infty. \tag{2.37}$$

The proof of Corollary 2.2 is now complete. □

3. The Subelliptic p-Laplacian

In this section we study the problem

$$\begin{cases} \frac{\partial u}{\partial t} = \mathbf{\Delta_p} u + V(w)u^{p-1} & \text{in} \quad \Omega \times (0,T), \quad 1 < p < 2, \\ u(w,0) = u_0(w) \geq 0 & \text{in} \quad \Omega, \\ u(w,t) = 0 & \text{on} \quad \partial\Omega \times (0,T). \end{cases} \tag{3.38}$$

Here we define the p-Laplacian in the weak sense, i.e., by considering the variational analysis of the p-energy form,

$$E(\phi) := \frac{1}{p}\Big\{\int_\Omega |\nabla_{\mathbb{H}^n}\phi(w)|^p dw - \int_\Omega V(w)|\phi(w)|^p dw\Big\}, \tag{3.39}$$

with $\phi \in C_c^\infty(\Omega)$ and V is a given real valued function on Ω. The nonlinear operator known as the p-Laplacian arises in the first variation of (3.2), which leads to the equation

$$-\mathbf{\Delta_p} u - V(w)u^{p-1} = D_\phi(E(\phi))|_{\phi=u} \tag{3.40}$$

where $\mathbf{\Delta_p} u = \nabla_{\mathbb{H}^n} \cdot (|\nabla_{\mathbb{H}^n} u|^{p-2}\nabla_{\mathbb{H}^n} u)$ and u is assumed to be nonnegative.

Let $\sigma^p_{\text{inf}}(V)$ denote the bottom of the normalized p-energy forms, which is defined by

$$\sigma^p_{\text{inf}}(V) := \inf_{0 \not\equiv \phi \in C_c^\infty(\Omega \backslash \mathcal{K})} \frac{\int_\Omega |\nabla_{\mathbb{H}^n}\phi|^p dw - \int_\Omega V|\phi|^p dw}{\int_\Omega |\phi|^p dw}.$$

We will show that nonexistence of positive solutions depends only on the size of the bottom of the normalized p-energy forms. As in the previous section, we consider some singular potentials (since they are critical) and we show that Hardy's constant is a cut-off point for the nonexistence of positive solutions.

The following Hardy type inequality on $\mathbb{H}^n$ is proved by Niu, Zhang, and Wang in [15].

Theorem 3.4. *Let $\phi \in C_0^\infty(\mathbb{H}^n \setminus \{0\})$, $1 < p < Q$. Then*

$$(\frac{Q-p}{p})^p \int_{\mathbb{H}^n} \frac{|z|^p}{(|z|^4+l^2)^{\frac{p}{2}}} |\phi|^p dzdl \leq \int_{\mathbb{H}^n} |\nabla_{\mathbb{H}^n}\phi|^p dzdl.$$

The following is the main result of this section.

Theorem 3.5. *Let $n \in \mathbb{N}$, $\frac{4n+4}{2n+3} \leq p < 2$ and $V \in L^1_{loc}(\Omega \setminus \mathcal{K})$ where $\mathcal{K}$ is a closed Lebesgue null subset of Ω. If*

$$\sigma^p_{\inf}((1-\epsilon)V) := \inf_{0 \not\equiv \phi \in C_c^\infty(\Omega \setminus \mathcal{K})} \frac{\int_\Omega |\nabla_{\mathbb{H}^N}\phi|^p - \int_\Omega (1-\epsilon)V|\phi|^p dw}{\int_\Omega |\phi|^p dw} = -\infty$$

for some $\epsilon > 0$, then (3.1) has no general positive local solution off of $\mathcal{K}$.

Proof: We argue by contradiction. Given any $T > 0$, let $u : [0,T) \longrightarrow L^1(\Omega)$ be a general positive local solution to (3.1) in $(\Omega \setminus \mathcal{K}) \times (0,T)$ with $u_0 \geq 0$ but not identically zero. Multiply both sides of (3.1) by the test function $|\phi|^p/u^{p-1}$ where $\phi \in C_c^\infty(\Omega \setminus \mathcal{K})$, and integrate over Ω, to get

$$\frac{1}{2-p}\frac{d}{dt}\int_\Omega u^{2-p}|\phi|^p dx - \int_\Omega \mathbf{\Delta}_\mathbf{p} u(\frac{|\phi|^p}{u^{p-1}})dw = \int_\Omega V|\phi|^p dw. \tag{3.41}$$

It follows from integration by parts that

$$\int_\Omega \mathbf{\Delta}_\mathbf{p} u(\frac{|\phi|^p}{u^{p-1}})dw = -\int_\Omega |\nabla_{\mathbb{H}^n}u|^{p-2}\nabla_{\mathbb{H}^n}u \cdot \nabla_{\mathbb{H}^n}(\frac{|\phi|^p}{u^{p-1}})dw. \tag{3.42}$$

Since

$$|\nabla_{\mathbb{H}^n}u|^{p-2}\nabla_{\mathbb{H}^n}u \cdot \nabla_{\mathbb{H}^n}(\frac{|\phi|^p}{u^{p-1}})$$

$$-p|\nabla_{\mathbb{H}^n}u|^{p-2}\frac{|\phi|^{p-1}}{u^{p-1}}\nabla_{\mathbb{H}^n}u \cdot \nabla_{\mathbb{H}^n}|\phi| - (p-1)\frac{|\phi|^p}{u^p}|\nabla_{\mathbb{H}^n}u|^p,$$

$$\begin{aligned}\int_\Omega \mathbf{\Delta}_\mathbf{p} u(\frac{|\phi|^p}{u^{p-1}})dw &= (p-1)\int_\Omega |\nabla_{\mathbb{H}^n}u|^p \frac{\phi^p}{u^p}dw \\ &\quad - p\int_\Omega |\nabla_{\mathbb{H}^n}u|^{p-2}\frac{\phi^{p-1}}{u^{p-1}}\nabla_{\mathbb{H}^n}u \cdot \nabla_{\mathbb{H}^n}|\phi|dw,\end{aligned} \tag{3.43}$$

and then we have

$$\begin{aligned}\int_\Omega \mathbf{\Delta}_\mathbf{p} u(\frac{|\phi|^p}{u^{p-1}})dw &\geq (p-1)\int_\Omega |\nabla_{\mathbb{H}^n}u|^p \frac{|\phi|^p}{u^p}dw \\ &\quad - p\int_\Omega |\nabla_{\mathbb{H}^n}u|^{p-1}|\nabla_{\mathbb{H}^n}\phi|\frac{\phi^{p-1}}{u^{p-1}}dw.\end{aligned} \tag{3.44}$$

Here we can use the following elementary inequality: Let $p > 1$ and $w_1 \neq w_2$ be two positive real numbers. Then

$$w_1^p - w_2^p - p w_2^{p-1}(w_1 - w_2) > 0;$$

it follows that

$$(p-1)w_2^p - p w_2^{p-1} w_1 > -w_1^p.$$

We can take $w_2 = |\frac{\phi}{u}\nabla_{\mathbb{H}^n} u|$, $w_1 = |\nabla_{\mathbb{H}^n}\phi|$; then we have

$$\begin{aligned} &(p-1)\int_\Omega |\nabla_{\mathbb{H}^n} u|^p \frac{|\phi|^p}{u^p} dw - p\int_\Omega |\nabla_{H^n} u|^{p-1} |\nabla_{\mathbb{H}^n}\phi| \frac{|\phi|^{p-1}}{u^{p-1}} dw \\ &\geq -\int_\Omega |\nabla_{\mathbb{H}^n}\phi|^p dw. \end{aligned} \tag{3.45}$$

Therefore

$$\int_\Omega \mathbf{\Delta_p} u (\frac{|\phi|^p}{u^{p-1}}) dw \geq -\int_\Omega |\nabla_{\mathbb{H}^n}\phi|^p dw. \tag{3.46}$$

Substituting (3.9) into (3.4) and integrating from t_1 to t_2, where $0 < t_1 < t_2 < T$, we obtain

$$\begin{aligned} &\int_\Omega V(w)\phi^p dw - \int_\Omega |\nabla_{\mathbb{H}^n}\phi|^p dw \\ &\leq \frac{1}{(2-p)(t_2-t_1)} \int_\Omega (u^{2-p}(w,t_2) - u^{2-p}(w,t_1))\phi^p dw. \end{aligned} \tag{3.47}$$

Using Jensen's inequality for concave functions, we obtain

$$\int_\Omega \Big(u^{(2-p)}(w,t_i)\Big)^{\frac{(2n+2)}{p}} dw \leq C(|\Omega|)\Big(\int_\Omega u(w,t_i)dw\Big)^{\frac{(2-p)(2n+2)}{p}} < \infty.$$

Here we use the fact that Ω is bounded, whence $|\Omega|$ is finite, and that $\frac{4n+4}{2n+3} \leq p$. Therefore

$$u^{2-p}(w,t_i) \in L^{\frac{2n+2}{p}}(\Omega).$$

By Proposition A.1, we have, for any given $\epsilon > 0$,

$$\begin{aligned} &\frac{1}{(2-p)(t_2-t_1)} \int_\Omega \Big(u^{2-p}(w,t_2) - u^{2-p}(w,t_1)\Big)|\phi|^p dw \\ &\leq \frac{\epsilon}{1-\epsilon}\int_\Omega |\nabla_{\mathbb{H}^n}\phi|^p dw + C(\epsilon)\int_\Omega |\phi|^2 dw. \end{aligned} \tag{3.48}$$

Substituting (3.11) into (3.10), we obtain

$$\begin{aligned} &\int_\Omega V(w)|\phi|^p dw - \int_\Omega |\nabla_{\mathbb{H}^n}\phi|^p dw \\ &\leq \frac{\epsilon}{1-\epsilon}\int_\Omega |\nabla_{\mathbb{H}^n}\phi|^p dw + C(\epsilon)\int_\Omega |\phi|^p dw. \end{aligned} \tag{3.49}$$

Therefore

$$\inf_{0\not\equiv\phi\in C_c^\infty(\Omega\setminus\mathcal{K})} \frac{\int_\Omega |\nabla\phi|^p dw - \int_\Omega (1-\epsilon)V(w)|\phi|^p dw}{\int_\Omega |\phi|^p dw} \geq -(1-\epsilon)C(\epsilon) > -\infty. \tag{3.50}$$

This contradicts our assumption. The proof of Theorem 3.1 is now complete. □

Singular Potentials

We now consider the following potentials. First we treat

$$V(z,l) = \frac{c|z|^p}{(|z|^4+l^2)^{\frac{p}{2}}}, \tag{3.51}$$

and later

$$V(z,l) = \frac{c|z|^p}{(|z|^4+l^2)^{\frac{p}{2}}} + \frac{\beta|z|^p}{(|z|^4+l^2)^{\frac{p}{2}}} \sin\left(\frac{1}{(|z|^4+l^2)^{\frac{\alpha}{4}}}\right) \tag{3.52}$$

where $c > 0$, $1 < p < 2$, $\beta \in \mathbb{R}\setminus\{0\}$ and $\alpha > 0$. We prove that the nonexistence of positive solutions only depends on the size of c.

Corollary 3.3. *Let $0 \in \Omega$, $C_*(n) = (\frac{2n+2-p}{p})^p$, and $V(z,l)$ be defined by (3.14). Then the problem (3.1) has no general positive local solution off of $\mathcal{K}$ if $c > C_*(n)$ and $\frac{4n+4}{2n+3} \leq p < 2$.*

Proof. Given $\epsilon > 0$, we define the radial function, $\phi \in C_c^1(\Omega)\bigcap W^{1,\infty}(\Omega)$, by

$$\phi(\rho) = \begin{cases} \epsilon^{(\frac{2n+2-p}{p})} & \text{if } 0 \leq \rho \leq \epsilon \\ \rho^{-(\frac{2n+2-p}{p})} & \text{if } \epsilon \leq \rho \leq 1 \\ 2-\rho & \text{if } 1 \leq \rho \leq 2 \\ 0 & \text{if } \rho \geq 2. \end{cases} \tag{3.53}$$

Without loss of generality we assume that $B_{\mathbb{H}^n}(0,2) = \{w \in \mathbb{H}^n : d(0,w) < 2\} \subset \Omega$.

We want to show that

$$\sigma_{\inf}^p(V;\phi) := \inf_{0\not\equiv\phi\in C_c^\infty(\Omega\setminus\mathcal{K})} \frac{\int_\Omega |\nabla_{\mathbb{H}^n}\phi|^p dw - \int_\Omega V|\phi|^p dw}{\int_\Omega |\phi|^p dw} = -\infty.$$

Lemma 2.1 gives

$$|\nabla_{\mathbb{H}^n}\phi| = \begin{cases} 0 & \text{if } 0 \leq \rho < \epsilon \text{ and } \rho > 2 \\ (\frac{2n+2-p}{p})\rho^{-\frac{2n+2}{p}}\frac{r}{\rho} & \text{if } \epsilon < \rho < 1 \\ \frac{r}{\rho} & \text{if } 1 < \rho < 2. \end{cases} \tag{3.54}$$

Using $r = \rho(\sin\theta)^{1/2}$ then we obtain

$$\begin{aligned}\int_\Omega |\nabla_{\mathbb{H}^n}\phi|^p dw &= \mu_n\Big[(\frac{2n+2-p}{p})^p \int_\epsilon^1 \frac{d\rho}{\rho} + \int_1^2 \rho^{2n+1} d\rho\Big] \\ &= \mu_n\Big[-(\frac{2n+2-p}{p})^p \log\epsilon + \frac{2^{2n+2}-1}{2n+2}\Big],\end{aligned} \tag{3.55}$$

where

$$\mu_n = \Gamma_n \int_0^\pi (\sin\theta)^{\frac{p+2n-2}{2}} d\theta.$$

For the next integral, we write

$$V(z,l) = \frac{c|z|^p}{(|z|^4+l^2)^{\frac{p}{2}}} = \frac{cr^p}{\rho^{2p}} = \frac{(\sin\theta)^{p/2}}{\rho^p}.$$

Then

$$\begin{aligned}\int_\Omega V(w)|\phi|^p dw = \mu_n\Big[\epsilon^{p-2n-2} \int_0^\epsilon \rho^{2n+1-p} d\rho + \int_\epsilon^1 \frac{d\rho}{\rho} \\ + \int_1^2 (2-\rho)^p \rho^{2n+1-p} d\rho\Big].\end{aligned}$$

Applying the Generalized Second Mean Value Theorem (Proposition A.2) for integrals to the last integral above, we obtain

$$\int_\Omega V(w)|\phi|^p dw = \mu_n\Big[\frac{1}{2n+2-p} - \log\epsilon + \frac{\xi^{2n+2-p}-1}{2n+2-p}\Big] \tag{3.56}$$

where $\xi \in (1,2)$.

Next,

$$\begin{aligned}\int_\Omega |\phi|^p dw = \Gamma_n\Big[\epsilon^{p-2n-2} \int_0^\epsilon \rho^{2n+1} d\rho + \int_\epsilon^1 \rho^{p-1} d\rho \\ + \int_1^2 (2-\rho)^p \rho^{2n+1} d\rho\Big]\end{aligned}.$$

Applying the Generalized Second Mean Value Theorem (Proposition A.2) to the last integral above, we obtain

$$\int_\Omega |\phi|^p dw = \Gamma_n\Big[\frac{\epsilon^p}{2n+2} + \frac{1-\epsilon^p}{p} + \frac{\eta^{2n+2}-1}{2n+2}\Big] \quad \text{where} \quad \eta \in (1,2). \tag{3.57}$$

Substituting (3.18), (3.19) and (3.20) into the Rayleigh quotient,

$$\begin{aligned}\mathcal{R} &= \frac{\int_\Omega |\nabla_{\mathbb{H}^N}\phi|^p dw - \int_\Omega V|\phi|^p dw}{\int_\Omega |\phi|^p dw} \\ &= \frac{\mu_n\Big[-(\frac{2n+2-p}{p})^p \log\epsilon + \frac{2^{2n+2}-1}{2n+2} - c(\frac{1}{2n+2-p} - \log\epsilon + \frac{\xi^{2n+2-p}-1}{2n+2-p})\Big]}{\Gamma_n\Big[\frac{\epsilon^p}{2n+2} + \frac{1-\epsilon^p}{p} + \frac{\eta^{2n+2}-1}{2n+2}\Big]} \\ &= \frac{\mu_n\Big[(c-(\frac{2n+2-p}{p})^p)\log\epsilon + \frac{2^{2n+2}-1}{2n+2} - c(\frac{1}{2n+2-p} + \frac{\xi^{2n+2-p}-1}{2n+2-p})\Big]}{\Gamma_n\Big[\frac{\epsilon^p}{2n+2} + \frac{1-\epsilon^p}{p} + \frac{\eta^{2n+2}-1}{2n+2}\Big]}.\end{aligned} \tag{3.58}$$

Then taking limits as $\epsilon \longrightarrow 0^+$, we find that the right-hand side of (3.21) can become negative and arbitrarily large in magnitude, i.e.,

$$\sigma_{\inf}^p(V) := \inf_{0 \not\equiv \phi \in C_c^\infty(\Omega \backslash \mathcal{K})} \frac{\int_\Omega |\nabla_{\mathbb{H}^n}\phi|^p dw - \int_\Omega V|\phi|^p dw}{\int_\Omega |\phi|^p dw} = -\infty.$$

The proof of Corollary 3.1 is now complete. □

Corollary 3.4. *Let* $0 \in \Omega$, $C_*(n) = (\frac{2n+2-p}{p})^p$, *and* $V(z,l)$ *is defined by (3.15). Then the problem (3.1) has no general positive local solution off of* $\mathcal{K}$ *if* $c > C_*(n)$ *and* $\frac{4n+4}{2n+3} \leq p < 2$.

Proof. We define the radial function, $\phi \in C_c^1(\Omega) \bigcap W^{1,\infty}(\Omega)$, by (3.16). We want to show that

$$\sigma_{\inf}^p(V) := \inf_{0 \not\equiv \phi \in C_c^\infty(\Omega \backslash \mathcal{K})} \frac{\int_\Omega |\nabla_{\mathbb{H}^n}\phi|^p dw - \int_\Omega V|\phi|^p dw}{\int_\Omega |\phi|^p dw} = -\infty.$$

We already computed most of the relevant integrals; see (3.18), (3.19) and (3.20).

For the next integral, we write

$$\begin{aligned} V(z,l) &= \frac{c|z|^p}{(|z|^4+l^2)^{\frac{p}{2}}} + \frac{\beta|z|^p}{(|z|^4+l^2)^{\frac{p}{2}}} \sin\left(\frac{1}{(|z|^4+l^2)^{\frac{\alpha}{4}}}\right) \\ &= \frac{cr^p}{\rho^{2p}} + \frac{\beta r^p}{\rho^{2p}} \sin(\frac{1}{\rho^\alpha}). \end{aligned} \tag{3.59}$$

We want to compute

$$\int_\Omega V(w)|\phi|^p dw.$$

We only need to compute the following integral,

$$I = \int_\Omega \frac{|z|^p}{(|z|^4+l^2)^{\frac{p}{2}}} \sin\left(\frac{1}{(|z|^4+l^2)^{\frac{\alpha}{4}}}\right)|\phi|^p dw. \tag{3.60}$$

Since $r = \rho(\sin\theta)^{1/2}$ we write

$$\begin{aligned} I &= \int_\Omega \frac{|z|^p}{(|z|^4+l^2)^{\frac{p}{2}}} \sin\left(\frac{1}{(|z|^4+l^2)^{\frac{\alpha}{4}}}\right)|\phi|^p dw \\ &= \mu_n \int_0^2 \rho^{2n+1-p} \sin(\frac{1}{\rho^\alpha})|\phi|^p d\rho. \end{aligned} \tag{3.61}$$

Let

$$A = \int_0^\epsilon \epsilon^{p-2n-2}\rho^{2n+1-p} \sin(\frac{1}{\rho^\alpha}),$$

$$B = \int_\epsilon^1 \frac{1}{\rho} \sin(\frac{1}{\rho^\alpha}) d\rho,$$

$$C = \int_1^2 (2-\rho)^p \rho^{2n+1-p} \sin(\frac{1}{\rho^\alpha}),$$

then $I = \mu_n(A + B + C)$. Using a change of variable,

$$A = \frac{1}{\alpha}\epsilon^{p-2n-2}\int_{\epsilon^{-\alpha}}^{\infty} s^{\frac{-1}{\alpha}(2n+2-p+\alpha)}\sin(s)ds,$$

the alternating series method [11] gives

$$|A| \le \frac{4}{\alpha}(1 + o(1)) \quad \text{as} \quad \epsilon \longrightarrow 0^+. \tag{3.62}$$

Using a change of variable,

$$B = (\frac{1}{\alpha}\int_1^{\epsilon^{-\alpha}} \frac{\sin(s)}{s}ds), \tag{3.63}$$

and clearly

$$|B| \le \frac{4}{\alpha}(1 + o(1)) \quad \text{as} \quad \epsilon \longrightarrow 0^+. \tag{3.64}$$

Next,

$$C = \int_1^2 (2-\rho)^p \rho^{2n+1-p}\sin(\frac{1}{\rho^\alpha}).$$

As an upper bound we obtain

$$|C| \le (\frac{2^{2n+2-p}-1}{2n+2-p}). \tag{3.65}$$

Therefore

$$|I| \le \mu_n\Big(\frac{8}{\alpha}(1+o(1)) + \frac{2^{2n+2-p}-1}{2n+2-p}\Big). \tag{3.66}$$

Substituting (3.18), (3.19), (3.20) and (3.29) into the Rayleigh quotient,

$$\begin{aligned}\mathcal{R} &= \frac{\int_\Omega |\nabla_{\mathbb{H}^n}\phi|^p dw - \int_\Omega V|\phi|^p dw}{\int_\Omega |\phi|^p dw} \\ &= \frac{\mathcal{A} - \mathcal{B}}{\mathcal{C}}\end{aligned} \tag{3.67}$$

where

$$\mathcal{A} = \mu_n\Big[(c - (\frac{2n+2-p}{p})^p)\log\epsilon + \frac{2^{2n+2}-1}{2n+2}\Big]$$

$$\mathcal{B} = \mu_n\Big[c(\frac{1}{2n+2-p} + \frac{\xi^{2n+2-p}-1}{2n+2-p}) - \beta(\frac{8}{\alpha}(1+o(1)) + \frac{2^{2n+2-p}-1}{2n+2-p})\Big]$$

$$\mathcal{C} = \Gamma_n\Big[\frac{\epsilon^p}{2n+2} + \frac{1-\epsilon^p}{p} + \frac{\eta^{2n+2}-1}{2n+2}\Big]$$

Then taking limits as $\epsilon \longrightarrow 0^+$, we find that the right-hand side of (3.30) can become negative and arbitrarily large in magnitude, i.e.,

$$\sigma^p_{\inf}(V) := \inf_{0\not\equiv\phi\in C_c^\infty(\Omega\setminus\mathcal{K})} \frac{\int_\Omega |\nabla_{\mathbb{H}^n}\phi|^p dw - \int_\Omega V|\phi|^p dw}{\int_\Omega |\phi|^p dw} = -\infty.$$

The proof of Corollary 3.2 is now complete. □

Remark. We would like to point out that all the new results above hold on the whole space $\mathbb{H}^n$, which is a consequence of the bounded domain case (see [13], [14]).

A

Proposition 1.1. *Let Ω be a bounded domain in $\mathbb{H}^n$ and $1 \leq p < Q$, $M \in L^{Q/p}(\Omega)$, $\phi \in W_0^{1,p}(\Omega)$. Then for each $\epsilon > 0$ there exists a positive constant $C(\epsilon)$ such that*

$$\int_\Omega M(w)|\phi|^p dw \leq \frac{\epsilon}{1-\epsilon}\int_\Omega |\nabla_{\mathbb{H}^n}\phi|^p dw + C(\epsilon)\int_\Omega |\phi|^p dw \tag{1.68}$$

Proof. Let us truncate $M(w)$ to $M_k(w) = min\{M(w), k\}$. Then $||M_k - M||_{L^{Q/p}} \longrightarrow 0$ as $k \longrightarrow \infty$. Clearly we have

$$\int_\Omega M|\phi|^p dw \leq \int_\Omega |M - M_k||\phi|^p dw + k\int_\Omega |\phi|^p dw.$$

Using Hölder's inequality for the first integral on the right side,

$$\int_\Omega M|\phi|^p dw \leq \Big(\int_\Omega |M - M_n|^{\frac{Q}{p}} dw\Big)^{\frac{p}{Q}}\Big(\int_\Omega |\phi|^{\frac{Qp}{Q-p}} dw\Big)^{\frac{Q-p}{Q}} + k\int_\Omega |\phi|^p dw.$$

It is well known that the following Sobolev inequality holds [7, 17] for $\phi \in C_0^\infty(\mathbb{H}^n)$:

$$\Big(\int_{\mathbb{H}^n} |\phi(w)|^q dw\Big)^{1/q} \leq C_{p,q}\Big(\int_{\mathbb{H}^n} |\nabla_{\mathbb{H}^n}\phi(w)|^p dw\Big)^{1/p}$$

provided that $1 \leq p < Q$ and $q = \frac{Qp}{Q-p}$. Therefore,

$$\int_\Omega M|\phi|^p dx \leq C_{p,q}\Big(\int_\Omega |M - M_n|^{\frac{Q}{p}} dx\Big)^{\frac{p}{Q}}\int_\Omega |\nabla_{\mathbb{H}^n}\phi|^p dw + k\int_\Omega |\phi|^p dw$$

which implies

$$\Big(\int_\Omega |M - M_n|^{\frac{Q}{p}} dx\Big)^{\frac{p}{Q}} \leq \frac{\epsilon}{(1-\epsilon)C_{p,q}}, \quad \text{for } k \text{ sufficiently large.}$$

Fix such an k; now let $k = C(\epsilon)$ we have desired inequality (A.1).

Proposition 1.2. *(Generalized Second Mean Value Theorem) If $f(x)$ and $g(x)$ are continuous in [a, b] and $g(x)$ is monotone increasing or monotone decreasing, there is a point ξ in (a,b) such that*

$$\int_a^b f(x)g(x)dx = g(a)\int_a^\xi f(x)dx + g(b)\int_\xi^b f(x)dx$$

Proof. This well known result appears in some of older elementary calculus texts. For a proof, see, for example, [11, p. 1191].

References

[1] J. A. Aguilar Crespo and I. Peral, Global behaviour of the Cauchy problem for some critical nonlinear parabolic equations, *SIAM J. Math. Anal*, **31** (2000), 1270-1294.

[2] P. Baras and J. A. Goldstein, The heat equation with a singular potential, *Trans. AMS*, **284** (1984), 121-139.

[3] R. Beals, B. Gaveau, P. Greiner, J. Vauthier, The Laguerre Calculus on the Heisenberg group: II, *Bull. Sci. Math.*, **110** (1986), 225-288.

[4] C. Berenstein, D. Chang, J. Tie, *Laguerre Calculus and Its Applications on the Heisenberg Group*, American Mathematical Society, 2001

[5] X. Cabré and Y. Martel, Existence versus explosion instantanée pour des équations de la chaleur linéaires avec potentiel singulier, *C. R. Acad. Sci. Paris*, **329** (1999), 973-978.

[6] Lorenzo D'Ambrosio,Critical degenerate inequalities on the Heisenberg group, *Manuscripta Math*, **106** 2001, 519-536.

[7] G. B. Folland, Subelliptic estimates and function spaces on nilpotent Lie groups, *Ark. Mat*, **13** (1975), 161-207.

[8] J. Garcia Azorero and I. Peral, Hardy inequalities and some critical elliptic and parabolic problems, *J. Diff. Equations*, **144** (1998), 441-476.

[9] N. Garofalo and E. Lanconelli, Frequency functions on the Heisenberg group, the uncertainty principle and unique continuation, *Ann. Inst. Fourier(Grenoble)* **40** (1990), 313-356

[10] J. A. Goldstein and I. Kombe, Instantaneous blow up, *Contemp. Math.* **327** (2003) 141-149.

[11] J. A. Goldstein and I. Kombe, Nonlinear parabolic differential equations with the singular lower order term, *Adv. Differential Equations* **10** (2003), 1153-1192.

[12] J. A. Goldstein and Q. S. Zhang, On a degenerate heat equation with a singular potential, *J. Functional Analysis*, **186** (2001), 342-359.

[13] J. A. Goldstein and Q. S. Zhang, Linear parabolic equations with strong singular potentials, *Trans. AMS*, **355** (2003), 197-211.

[14] I. Kombe, The linear heat equation with a highly singular, oscillating potential, *Proc. Amer. Math. Soc.* **132** (2004), 2683-2691.

[15] P. Niu, H. Zhang, Y. Wang, Hardy type and Rellich type inequalities on the Heisenberg group, *Proc. Amer. Math. Soc*, **129** (2001), 3623-3630

[16] E. Stein, *Harmonic Analysis, Real-Variable Methods, Orthgonality, and Oscillatory Integrals*, Princeton University Press, Princeton, NJ.

[17] N. Varopoulos, L. Saloff-Coste, T. Coulhon, *Analysis and Geometry on Groups*. Cambridge University Press, Cambridge, 1992

In: Focus on Evolution Equations
Editor: Gaston M. N'Guerekata
ISBN: 978-1-60021-342-7

Chapter 2

REMARKS ON THE SCHRÖDINGER EQUATION*

Robert Carroll†
University of Illinois, Urbana, IL, U. S. A.

Abstract

Various origins of linear and nonlinear Schrödinger equations are discussed in connection with diffusion, hydrodynamics, and fractal structure. The treatment is mainly expository, emphasizing the quantum potential, with a few new observations.

1. Introduction

Perhaps no subject has been the focus of as much mystery as "classical" quantum mechanics (QM) even though the standard Hilbert space framework provides an eminently satisfactory vehicle for determining accurate conclusions in many situations. This and other classical viewpoints provide also seven decimal place accuracy in QED for example. So why all the fuss? The erection of the Hilbert space edifice and the subsequent development of operator algebras (extending now into noncommutative (NC) geometry) has an air of magic. It works but exactly why it works and what it really represents remain shrouded in ambiguity. Also geometrical connections of QM and classical mechanics (CM) are still a source of new work and a modern paradigm focuses on the emergence of CM from QM (or below). Below could mean here a micro structure of space time (quantum foam, Cantorian spacetime, etc.). In addition there are beautiful stochastic theories for diffusion and QM. In terms of background information in book form we mention here e.g. [4, 6, 11, 12, 31, 27, 28, 31, 65, 74, 77, 79, 82, 84, 83, 86, 90, 92, 111, 114, 128, 134, 136] (the lecture notes [13, 14, 15, 16, 17] in a more polished and organized form should also eventually become part of a book in preparation). The present paper focuses on certain aspects of the Schrödinger equation (SE) involving the wave function form $\psi = Rexp(iS/\hbar)$, hydrodynamical versions, diffusion processes, quantum potentials, and fractal methods. The aim is to envision "structure", both mathematical and physical, and we avoid detailed technical discussion of mathematical fine points (cf. [27, 28, 32, 36, 91, 117, 134, 138] for various delicate matters). Rather than

*Communicated by Robert Gilbert

†E-mail address: rcarroll@math.uiuc.edu

looking at such matters as Markov processes with jumps for example we prefer to seek "meaning" for the Schrödinger equation via microstructure and fractals in connection with diffusion processes and kinetic theory.

2. Background for the Schrödinger Equation

First consider the SE in the form $(\mathbf{A1})\quad -(\hbar^2/2m)\psi'' + V\psi = i\hbar\psi_t$ so that for $\psi = Rexp(iS/\hbar)$ one obtains

$$S_t + \frac{S_X^2}{2m} + V - \frac{\hbar^2 R''}{2mR} = 0;\ \partial_t(R^2) + \frac{1}{m}(R^2 S')' = 0 \tag{2.1}$$

where $S' \sim \partial S/\partial X$. Writing $P = R^2$ (probability density $\sim |\psi|^2$) and $Q = -(\hbar^2/2m)(R''/R)$ (quantum potential) this becomes

$$S_t + \frac{(S')^2}{2m} + Q + V = 0;\ P_t + \frac{1}{m}(PS')' = 0 \tag{2.2}$$

and this has some hydrodynamical interpretations in the spirit of Madelung. Indeed going to [41] for example we take $p = S'$ with $p = m\dot{q}$ for $\dot{q}$ a velocity (or "collective" velocity - unspecified). Then (2.2) can be written as ($\rho = mP$ is an unspecified mass density)

$$S_t + \frac{p^2}{2m} + Q + V = 0;\ P_t + \frac{1}{m}(Pp)' = 0;\ p = S';\ P = R^2;\ Q = -\frac{\hbar^2}{2m}\frac{R''}{R} = -\frac{\hbar^2}{2m}\frac{\partial^2\sqrt{\rho}}{\sqrt{\rho}} \tag{2.3}$$

Note here

$$\frac{\partial^2\sqrt{\rho}}{\sqrt{\rho}} = \frac{1}{4}\left[\frac{2\rho''}{\rho} - \left(\frac{\rho'}{\rho}\right)^2\right] \tag{2.4}$$

Now from $S' = p = m\dot{q} = mv$ one has

$$P_t + (P\dot{q})' = 0 \equiv \rho_t + (\rho\dot{q})' = 0;\ S_t + \frac{p^2}{2m} + V - \frac{\hbar^2}{2m}\frac{\partial^2\sqrt{\rho}}{\sqrt{\rho}} = 0 \tag{2.5}$$

Differentiating the second equation in X yields ($\partial \sim \partial/\partial X,\ v = \dot{}\,)q$

$$mv_t + mvv' + \partial V - \frac{\hbar^2}{2m}\partial\left(\frac{\partial\sqrt{\rho}}{\sqrt{\rho}}\right) = 0 \tag{2.6}$$

Consequently, multiplying by $p = mv$ and ρ respectively in (2.5) and (2.6), we obtain

$$m\rho v_t + m\rho vv' + \rho\partial V - \frac{\hbar^2}{2m}\rho\partial\left(\frac{\partial^2\sqrt{\rho}}{\sqrt{\rho}}\right) = 0;\ mv\rho_t + mv(\rho' v + \rho v') = 0 \tag{2.7}$$

Then adding in (2.7) we get

$$\partial_t(\rho v) + \partial(\rho v^2) + \frac{\rho}{m}\partial V - \frac{\hbar^2}{2m^2}\rho\partial\left(\frac{\partial^2\sqrt{\rho}}{\sqrt{\rho}}\right) = 0 \tag{2.8}$$

This is similar to an equation in [41] (called an "Euler" equation) and it definitely has a hydrodynamic flavor (cf. also [60]).

Now go to [124] and write (2.6) in the form ($mv = p = S'$)

$$\frac{\partial v}{\partial t} + (v \cdot \nabla)v = -\frac{1}{m}\nabla(V+Q);\ v_t + vv' = -(1/m)\partial(v+Q) \quad (2.9)$$

The higher dimensional form is not considered here but matters are similar there. This equation (and (2.8)) is incomplete as a hydrodynamical equation as a consequence of a missing term $-\rho^{-1}\nabla\mathfrak{p}$ where $\mathfrak{p}$ is the pressure (cf. [81]). Hence one "completes" the equation in the form

$$m\left(\frac{\partial v}{\partial t} + (v \cdot \nabla)v\right) = -\nabla(V+Q) - \nabla F;\ mv_t + mvv' = -\partial(V+Q) - F' \quad (2.10)$$

where $(\mathbf{A2})$ $\nabla F = (1/R^2)\nabla\mathfrak{p}$ (or $F' = (1/R^2)\mathfrak{p}'$). By the derivations above this would then correspond to an extended SE of the form

$$i\hbar\frac{\partial\psi}{\partial t} = -\frac{\hbar^2}{2m}\Delta\psi + V\psi + F\psi \quad (2.11)$$

provided one can determine F in terms of the wave function ψ. One notes that it a necessary condition here involves $curlgrad(F) = 0$ or $(\mathbf{A3})$ $curl(R^{-2}\nabla\mathfrak{p}) = 0$ which enables one to take e.g. $(\mathbf{A4})$ $\mathfrak{p} = -bR^2 = -b|\psi|^2$. For one dimension one writes $(\mathbf{A5})$ $F' = -b(1/R^2)\partial|\psi|^2 = -(2bR'/R) \Rightarrow F = -2blog(R) = -blog(|\psi|^2)$. Consequently one has a corresponding SE

$$i\hbar\frac{\partial\psi}{\partial t} = -\frac{\hbar^2}{2m}\psi'' + V\psi - b(log|\psi|^2)\psi \quad (2.12)$$

This equation has a number of nice features discussed in [124] (but serious drawbacks as indicated in [23] - cf. also [37, 40, 42, 43, 56, 107, 108, 109]). For example $(\mathbf{A6})$ $\psi = \beta G(x-vt)exp(ikx - i\omega t)$ is a solution of (2.12) with $V = 0$ and for $v = \hbar k/m$ one gets $(\mathbf{A7})$ $\psi = cexp[-(B/4)(x - vt + d)^2]exp(ikx - i\omega t)$ where $B = 4mb/\hbar^2$. Normalization $\int_{-\infty}^{\infty}|\psi|^2 = 1$ is possible with $(\mathbf{A8})$ $|\psi|^2 = \delta_m(\xi) = \sqrt{m\alpha/\pi}exp(-\alpha m\xi^2)$ where $\alpha = 2b/\hbar^2$, $d = 0$, and $\xi = x - vt$. For $m \to \infty$ we see that δ_m becomes a Dirac delta and this means that motion of a particle with big mass is strongly localized. This is impossible for ordinary QM since $exp(ikx - i\omega t)$ cannot be localized as $m \to \infty$. Such behavior helps to explain the so-called collapse of the wave function and since superposition does not hold Schrödinger's cat is either dead or alive. Further $v = k\hbar/m$ is equivalent to the deBroglie relation $\lambda = h/p$ since $\lambda = (2\pi/k) = 2\pi(\hbar/mv) = 2\pi(h/2\pi)(1/p)$.

Remark 2.1. We go now to [70] and the linear SE in the form $(\mathbf{A9})$ $i(\partial\psi/\partial t) = -(1/2m)\Delta\psi + U(\vec{r})\psi$; such a situation leads to the Ehrenfest equations which have the form $(\mathbf{A10})$ $<\vec{v}>= (d/dt) <\vec{r}>$ and $<\vec{r}>= \int d^3x|\psi(\vec{r},t)|^2\vec{r}$ and $(\mathbf{A11})$ $m(d/dt) <\vec{v}>= \vec{F}(t)$ with $\vec{F}(t) = -\int d^3x|\psi(\vec{r},t)|^2\vec{\nabla}U(\vec{r})$. Thus the quantum expectation values of position and velocity of a suitable quantum system obey the classical equations of motion and the amplitude squared is a natural probability weight.

The result tells us that besides the statistical fluctuations quantum systems posess an extra source of indeterminacy, regulated in a very definite manner by the complex wave function. The Ehrenfest theorem can be extended to many point particle systems and in [70] one singles out the kind of nonlinearities that violate the Ehrenfest theorem. A theorem is proved that connects Galilean invariance, and the existence of a Lagrangian whose Euler-Lagrange equation is the SE, to the fulfillment of the Ehrenfest theorem. ■

Remark 2.2. There are many problems with the quantum mechanical theory of derived nonlinear SE (NLSE) but many examples of realistic NLSE arise in the study of superconductivity, Bose-Einstein condensates, stochastic models of quantum fluids, etc. and the subject demands further study. We make no attempt to survey this here but will give an interesting example later from [23] related to fractal structures where a number of the difficulties are resolved. For further information on NLSE, in addition to the references above, we refer to [7, 38, 54, 56, 70, 71, 72, 129, 130, 131, 140, 141] for some typical situations (the list is not at all complete and we apologize for omissions). Let us mention a few cases.

- The program of [70] introduces a Schrödinger Lagrangian for a free particle including self-interactions of any nonlinear nature but no explicit dependence on the space of time coordinates. The corresponding action is then invariant under spatial coordinate transformations and by Noether's theorem there arises a conserved current and the physical law of conservation of linear momentum. The Lagrangian is also required to be a real scalar depending on the phase of the wave function only through its derivatives. Phase transformations will then induce the law of conservation of probability identified as the modulus squared of the wave function. Galilean invariance of the Lagrangian then determines a connection betwee the probability current and the linear momentum which insures the validity of the Ehrenfest theorem.

- We turn next to [72] for a statistical origin for QM (cf. also [11, 38, 71, 73, 111, 118, 133]). The idea is to build a program in which the microscopic motion, underlying QM, is described by a rigorous dynamics different from Brownian motion (thus avoiding unnecessary assumptions about the Brownian nature of the underlying dynamics). The Madelung approach gives rise to fluid dynamical type equations with a quantum potential, the latter being capable of interpretation in terms of a stress tensor of a quantum fluid. Thus one shows in [72] that the quantum state corresponds to a subquantum statistical ensemble whose time evolution is governed by classical kinetics in the phase space. The equations take the form

$$\rho_t + \partial_x(\rho u) = 0;\ \partial_t(\mu\rho u_i) + \partial_j(\rho\phi_{ij}) + \rho\partial_{x_i}V = 0;\ \partial_t(\rho E) + \partial_x(\rho S) - \rho\partial_t V = 0 \tag{2.13}$$

$$\frac{\partial S}{\partial t} + \frac{1}{2\mu}\left(\frac{\partial S}{\partial x}\right)^2 + \mathcal{W} + V = 0 \tag{2.14}$$

for two scalar fields ρ, S determining a quantum fluid. These can be rewritten as

$$\frac{\partial \xi}{\partial t} + \frac{1}{\mu}\frac{\partial^2 S}{\partial x^2} + \frac{1}{\mu}\frac{\partial \xi}{\partial x}\frac{\partial S}{\partial x} = 0; \tag{2.15}$$

$$\frac{\partial S}{\partial t} - \frac{\eta^2}{4\mu}\frac{\partial^2 \xi}{\partial x^2} - \frac{\eta^2}{8\mu}\left(\frac{\partial \xi}{\partial x}\right)^2 + \frac{1}{2\mu}\left(\frac{\partial S}{\partial x}\right)^2 + V = 0$$

where $\xi = log(\rho)$ and for $\Omega = (\xi/2) + (i/\eta)S = log\Psi$ with $m = N\mu,\ \mathcal{V} = NV$, and $\hbar = N\eta$ one arrives at a SE

$$i\hbar\frac{\partial\Psi}{\partial t} = -\frac{\hbar^2}{2m}\frac{\partial^2\Psi}{\partial x^2} + \mathcal{V}\Psi \tag{2.16}$$

Further one can write $\Psi = \rho^{1/2}exp(i\mathfrak{S}/\hbar)$ with $\mathfrak{S} = NS$ and here $N = \int|\Psi|^2d^nx$. The analysis is very interesting. ■

Remark 2.3. Now in [44] one is obliged to use the form $\psi = Rexp(iS/\hbar)$ to make sense out of the constructions (this is no problem with suitable provisos, e.g. that S is not constant - cf. [8, 11, 47, 48]). Thus note from $(\mathbf{A12})$ $\psi'/\psi = (R'/R) + i(S'/\hbar)$ with $\Im(\psi'/\psi) = (1/m)S' \sim p/m$ (see also (2.19) below). Also note $(\mathbf{A13})$ $J = (\hbar/m)\Im\psi^*\psi'$ and $\rho = R^2 = |\psi|^2$ represent a current and a density respectively. Then using $p = mv = m\dot{q}$ one can write $(\mathbf{A14})$ $v = (\hbar/m)\Im(\psi'/\psi)$ and $J = (\hbar/m)\Im|\psi|^2(\psi^*\psi'/|\psi|^2) = (\hbar/m)\Im(\rho v)$. Then look at the SE in the form $i\hbar\psi_t = -(\hbar^2/2m)\psi'' + V\psi$ with $\psi_t = (R_t + iS_tR/\hbar)exp(iS/\hbar)$ and $\psi_{xx} = [(R' + (iS'R/\hbar)exp(iS/\hbar)]' = [R'' + (2iS'R'/\hbar) + (iS''R/\hbar) + (iS'/\hbar)^2R]exp(iS/\hbar)$ which means

$$-\frac{\hbar^2}{2m}\left[R'' - \left(\frac{S'}{\hbar}\right)^2 + \frac{2iS'r'}{\hbar} + \frac{iS''R}{\hbar}\right] + VR = i\hbar\left[R_t + \frac{iS_tR}{\hbar}\right] \Rightarrow \tag{2.17}$$

$$\Rightarrow \partial_t R^2 + \frac{1}{m}(R^2S')' = 0;\ S_t + \frac{(S')^2}{2mR} - \frac{\hbar^2R''}{2mR} + V = 0$$

This can also be written as

$$\partial_t\rho + \frac{1}{m}\partial(p\rho) = 0;\ S_t + \frac{p^2}{2m} + Q + V = 0 \tag{2.18}$$

where $Q = -\hbar^2R''/2mR$. Now we sketch the philosophy of [44, 45] in part. Most of such aspects are omitted and we try to isolate the essential mathematical features. First one emphasizes configurations based on coordinates whose motion is choreographed by the SE according to the rule (1-D only here)

$$\dot{q} = v = \frac{\hbar}{m}\Im\frac{\psi^*\psi'}{|\psi|^2} \tag{2.19}$$

where $(\mathbf{A15})$ $i\hbar\psi_t = -(\hbar^2/2m)\psi'' + V\psi$. The argument for (2.19) is based on obtaining the simplest Galilean and time reversal invariant form for velocity, transforming correctly under velocity boosts. This leads directly to (2.19) ($\sim$ $(\mathbf{A14})$) so that Bohmian mechanics (BM) is governed by (2.19) and **(A15)**. It's a fairly convincing argument and no recourse to Floydian time seems possible (cf. [11, 48, 50, 51]). Note however that if $S = c$ then $\dot{q} = v = (\hbar/m)\Im(R'/R) = 0$ while $p = S' = 0$ so perhaps this formulation avoids the $S = 0$ problems indicated in [11, 48, 50, 51]. One notes also that BM depends only on the Riemannian structure $g = (g_{ij}) = (m_i\delta_{ij})$ in the form $(\mathbf{A16})$ $\dot{q} = \hbar\Im(grad\psi/\psi);\ i\hbar\psi_t = -(\hbar^2/2)\Delta\psi + V\psi$. What makes the constant $\hbar/m$ in (2.19) important here is that with this value the probability density $|\psi|^2$ on configuration space is equivariant. This means that via the evolution of probability densities

$\rho_t + div(v\rho) = 0$ (as in (2.18) with $v \sim p/m$) the density $\rho = |\psi|^2$ is stationary relative to ψ, i.e. $\rho(t)$ retains the form $|\psi(q,t)|^2$. One calls $\rho = |\psi|^2$ the quantum equilibrium density (QED) and says that a system is in quantum equilibrium when its coordinates are randomly distributed according to the QED. The quantum equilibrium hypothesis (QHP) is the assertion that when a system has wave function ψ the distribution ρ of its coordinates satisfies $\rho = |\psi|^2$. ■

Remark 2.4. We extract here from [61, 62, 63] (cf. also the references there for background and [52, 53, 68] for some information geometry). There are a number of interesting results connecting uncertainty, Fisher information, and QM and we make no attempt to survey the matter. Thus first recall that the classical Fisher information associated with translations of a 1-D observable X with probability density $P(x)$ is

$$F_X = \int dx\, P(x)([log(P(x)]')^2 > 0 \tag{2.20}$$

One has a well known Cramer-Rao inequality $(\mathbf{A17})$ $Var(X) \geq F_X^{-1}$ where $Var(X) \sim$ variance of X. A Fisher length for X is defined via $(\mathbf{A18})$ $\delta X = F_X^{-1/2}$ and this quantifies the length scale over which $p(x)$ (or better $log(p(x))$) varies appreciably. Then the root mean square deviation ΔX satisfies $(\mathbf{A19})$ $\Delta X \geq \delta X$. Let now P be the momentum observable conjugate to X, and P_{cl} a classical momentum observable corresponding to the state ψ given via $(\mathbf{A20})$ $p_{cl}(x) = (\hbar/2i)[(\psi'/\psi) - (\bar{\psi}'/\bar{\psi})]$ (cf. (2.19)). One has the identity $(\mathbf{A21})$ $< p >_\psi = < p_{cl} >_\psi$ following from **(A20)** with integration by parts. Now define the nonclassical momentum by $p_{nc} = p - p_{cl}$ and one shows then $(\mathbf{A21})$ $\Delta X \Delta p \geq \delta X \Delta p \geq \delta X \Delta p_{nc} = \hbar/2$. Now go to [62] now where two proofs are given for the derivation of the SE from the exact uncertainty principle (as in **(A21)**). Thus consider a classical ensemble of n-dimensional particles of mass m moving under a potential V. The motion can be described via the HJ and continuity equations

$$\frac{\partial s}{\partial t} + \frac{1}{2m}|\nabla s|^2 + V = 0;\ \frac{\partial P}{\partial t} + \nabla \cdot \left[P\frac{\nabla s}{m}\right] = 0 \tag{2.21}$$

for the momentum potential s and the position probability density P (note that we have interchanged p and P from [62] - note also there is no quantum potential and this will be supplied by the information term). These equations follow from the variational principle $\delta L = 0$ with Lagrangian

$$L = \int dt\, d^n x\, P\left[\frac{\partial s}{\partial t} + \frac{1}{2m}|\nabla s|^2 + V\right] \tag{2.22}$$

It is now assumed that the classical Lagrangian must be modified due to the existence of random momentum fluctuations. The nature of such fluctuations is immaterial for (cf. [62] for discussion) and one can assume that the momentum associated with position x is given by $(\mathbf{A22})$ $p = \nabla s + N$ where the fluctuation term N vanishes on average at each point x. Thus s changes to being an average momentum potential. It follows that the average kinetic energy $< |\nabla s|^2 > /2m$ appearing in (2.22) should be replaced by $< |\nabla s + N|^2 > /2m$ giving rise to

$$L' = L + (2m)^{-1}\int dt < N \cdot N >= L + (2m)^{-1}\int dt (\Delta N)^2 \tag{2.23}$$

where $\Delta N = < N \cdot N >^{1/2}$ is a measure of the strength of the fluctuations. The additional term is specified uniquely, up to a multiplicative constant, by the following three assumptions

1. Action principle: L' is a scalar Lagrangian with respect to the fields P and s where the principle $\delta L' = 0$ yields causal equations of motion. Thus $(\mathbf{A23})$ $(\Delta N)^2 = \int d^n x\, pf(P, \nabla P, \partial P/\partial t, s, \nabla s, \partial s/\partial t, x, t)$ for some scalar function f.

2. Additivity: If the system comprises two independent noninteracting subsystems with $P = P_1 P_2$ then the Lagrangian decomposes into additive subsystem contributions; thus $(\mathbf{A24})$ $f = f_1 + f_2$ for $P = P_1 P_2$.

3. Exact uncertainty: The strength of the momentum fluctuation at any given time is determined by and scales inversely with the uncertainty in position at that time. Thus $(\mathbf{A25})$ $\Delta N \to k\Delta N$ for $x \to x/k$. Moreover since position uncertainty is entirely characterized by the probability density P at any given time the function f cannot depend on s, nor explicitly on t, nor on $\partial P/\partial t$.

The following theorem is then asserted (see [13, 62] for the proofs).

Theorem 2..1. The above 3 assumptions imply $(\mathbf{A26})$ $(\Delta N)^2 = c \int d^n x\, P|\nabla log(P)|^2$ where c is a positive universal constant.

Corollary 2..1. It follows from (2.23) that the equations of motion for p and s corresponding to the principle $\delta L' = 0$ are

$$i\hbar\frac{\partial\psi}{\partial t} = -\frac{\hbar^2}{2m}\nabla^2\psi + V\psi \tag{2.24}$$

where $\hbar = 2\sqrt{c}$ and $\psi = \sqrt{P}exp(is/\hbar)$. ■

Remark 2.5. We sketch here for simplicity and clarity another derivation of the SE along similar ideas following [132]. Let $P(y^i)$ be a probability density and $P(y^i + \Delta y^i)$ be the density resulting from a small change in the y^i. Calculate the cross entropy via

$$J(P(y^i + \Delta y^i) : P(y^i)) = \int P(y^i + \Delta y^i) log\frac{P(y^i + \Delta y^i)}{P(y^i)} d^n y \simeq \tag{2.25}$$

$$\simeq \left[\frac{1}{2}\int \frac{1}{P(y^i)}\frac{\partial P(y^i)}{\partial y^i}\frac{\partial P(y^i)}{\partial y^k)} d^n y\right] \Delta y^i \Delta y^k - I_{jk}\Delta y^i \Delta y^k$$

The I_{jk} are the elements of the Fisher information matrix. The most general expression has the form

$$I_{jk}(\theta^i) = \frac{1}{2}\int \frac{1}{P(x^i|\theta^i)}\frac{\partial P(x^i|\theta^i)}{\partial \theta^j}\frac{\partial P(x^i|\theta^i)}{\partial \theta^k} d^n x \tag{2.26}$$

where $P(x^i|\theta^i)$ is a probability distribution depending on parameters θ^i in addition to the x^i. For $(\mathbf{A27})$ $P(x^i|\theta^i) = P(x^i + \theta^i)$ one recovers (2.25) (straightforward - cf. [132]). If P is defined over an n-dimensional manifold with positive inverse metric g^{ik} one obtains a natural definition of the information associated with P via

$$I = g^{ik} I_{ik} = \frac{g^{ik}}{2}\int \frac{1}{P}\frac{\partial P}{\partial y^i}\frac{\partial P}{\partial y^k} d^n y \tag{2.27}$$

Now in the HJ formulation of classical mechanics the equation of motion takes the form

$$\frac{\partial S}{\partial t}+\frac{1}{2}g^{\mu\nu}\frac{\partial S}{\partial x^\mu}\frac{\partial S}{\partial x^\nu}+V=0 \tag{2.28}$$

where $g^{\mu\nu}=diag(1/m,\cdots,1/m)$. The velocity field u^μ is given by $(\mathbf{A28})$ $u^\mu=g^{\mu\nu}(\partial S/\partial x^\nu)$. When the exact coordinates are unknown one can describe the system by means of a probability density $P(t,x^\mu$ with $(\mathbf{A29})$ $\int Pd^nx=1$ and $(\mathbf{A30})$ $(\partial P/\partial t)+(\partial/\partial x^\mu)(Pg^{\mu\nu}(\partial S/\partial x^\nu)=0$. These equations completely describe the motion and can be derived from the Lagrangian

$$L_{CL}=\int P\left\{\frac{\partial S}{\partial t}+\frac{1}{2}g^{\mu\nu}\frac{\partial S}{\partial x^\mu}\frac{\partial S}{\partial x^\nu}+V\right\}dtd^nx \tag{2.29}$$

using fixed endpoint variation in S and P. Quantization is obtained by adding a term proportional to the information I defined in (2.27). This leads to

$$L_{QM}=L_{CL}+\lambda I=\int P\left\{\frac{\partial S}{\partial t}+\frac{1}{2}g^{\mu\nu}\left[\frac{\partial S}{\partial x^\mu}\frac{\partial S}{\partial x^\nu}+\frac{\lambda}{P^2}\frac{\partial P}{\partial x^\mu}\frac{\partial P}{\partial x^\nu}\right]+V\right\}dtd^nx \tag{2.30}$$

Fixed endpoint variation in S leads again to **(A30)** while variation in P leads to

$$\frac{\partial S}{\partial t}+\frac{1}{2}g^{\mu\nu}\left[\frac{\partial S}{\partial x^\mu}\frac{\partial S}{\partial x^\nu}+\lambda\left(\frac{1}{P^2}\frac{\partial P}{\partial x^\mu}\frac{\partial P}{\partial x^\nu}-\frac{2}{P}\frac{\partial^2 P}{\partial x^\mu\partial x^\nu}\right)\right]+V=0 \tag{2.31}$$

These equations are equivalent to the Schrödinger equation if $(\mathbf{A31})$ $\psi=\sqrt{P}exp(iS/\hbar)$ with $\lambda=(2\hbar)^2$ (cf. Section 6). ■

Remark 2.6. The SE gives to a probability distribution $\rho=|\psi|^2$ (with suitable normalization) and to this one can associate an information entropy $S(t)$ (actually configuration information entropy) $(\mathbf{A32})$ $S=-\int \rho log(\rho)d^3x$ which is typically not a conserved quantity (S is an unfortunate notation here but we retain it momentarily since no confusion should arise). The rate of change in time of S can be readily found by using the continuity equation $(\mathbf{A33})$ $\partial_t\rho=-\nabla\cdot(v\rho)$ where v is a current velocity field Note here (cf. also [126])

$$\frac{\partial S}{\partial t}=-\int\rho_t(1+log(\rho))dx=\int(1+log(\rho))\partial(v\rho) \tag{2.32}$$

Note that a formal substitution of $v=-u$ in **(A33)** implies the standard free Browian motion outcome $(\mathbf{A34})$ $dS/dt=D\cdot\int[(\nabla\rho)^2/\rho)d^3x=D\cdot Tr\mathfrak{F}\geq 0$ - use $(\mathbf{A35})$ $u=D\nabla log(\rho)$ with $D=\hbar/2m)$ and (2.32) with $\int(1+log(\rho))\partial(v\rho)=-\int v\rho\partial log(\rho)=-\int v\rho'\sim\int((\rho')^2/\rho)$ modulo constants involving D etc. Recall here $mfF\sim-(2/D^2)\int\rho Qdx=\int dx[(\nabla\rho)^2/\rho]$ is a functional form of Fisher information. A high rate of information entropy production corresponds to a rapid spreading (flattening down) of the probablity density. This delocalization feature is concomitant with the decay in time property quantifying the time rate at which the far from equilibrium system approaches its stationary state of equilibrium $(\mathbf{A36})$ $d/dtTr\mathfrak{F}\leq 0$. ■

Remark 2.7. Now going back to the quantum context one admits general forms of the current velocity v. For example consider a gradient field $v = b - u$ where the so-called forward drift $b(x,t)$ of the stochastic process depends on a particular diffusion model. Then one can rewrite the continuity equation as a standard Fokker-Plank equation (**A37**) $\partial_t\rho = D\Delta\rho - \nabla\cdot(b\rho)$. Boundary restrictions requiring ρ, $v\rho$, and $b\rho$ to vanish at spatial infinities or at boundaries yield the general entropy balance equation

$$\frac{dS}{dt} = \int\left[\rho(\nabla\cdot b) + D\cdot\frac{(\nabla\rho)^2}{\rho}\right]d^3x \equiv -D\frac{dS}{dt} = \int\rho(v\cdot u)d^3x =< v\cdot u > \quad (2.33)$$

The first term in the first equation is not positive definite and can be interpreted as an entropy flux while the second term refers to the entropy production proper. The flux term represents the mean value of the drift field divergence $\nabla\cdot b$ which by itself is a local measure of the flux incoming to or outgoing from an infinitesimal surrounding of x at time t. If locally $(\nabla\cdot b)(x,t) > 0$ on an infinitesimal time scale we would encounter a local entropy increase in the system (increasing disorder) while in case $(\nabla\cdot b)(x,t) < 0$ one thinks of local entropy loss or restoration or order. Only in the situation $< \nabla\cdot b >= 0$ is there no entropy production. Quantum dynamics permits more complicated behavior. One looks first for a general criterion under which the information entropy **(A32)** is a conserved quantity. Consider (2.8) and invoke the diffusion current to write (recall $u = D(\nabla\rho)/\rho$)

$$D\frac{dS}{dt} = -\int[\rho^{-1/2}(\rho v)]\cdot[\rho^{-1/2}(D\nabla\rho)]d^3x \quad (2.34)$$

Then by means of the Schwarz inequality one has (**A38**) $D|dS/dt| \leq< v^2 >^{1/2}< u^2 >^{1/2}$ so a necessary (but insufficient) condition for $dS/dt \neq 0$ is that both $< v^2 >$ and $< u^2 >$ are nonvanishing. On the other hand a sufficient condition for $dS/dt = 0$ is that either one of these terms vanishes. Indeed in view of (**A39**) $< u^2 >= D^2\int[(\nabla\rho)^2/\rho]d^3x$ the vanishing information entropy production implies $dS/dt = 0$; the vanishing diffusion current does the same job. ∎

Remark 2.8. We develop a little more perspective now (following [55] - first paper). Recall Q written out as

$$Q = 2D^2\frac{\Delta\rho^{1/2}}{\rho^{1/2}} = D^2\left[\frac{\Delta\rho}{\rho} - \frac{1}{2\rho^2}(\nabla\rho)^2\right] = \frac{1}{2}u^2 + D\nabla\cdot u \quad (2.35)$$

where $u = D\nabla log(\rho)$ is called an osmotic velocity field. The standard Brownian motion involves $v = -u$, known as the diffusion current velocity and (up to a dimensional factor) is identified with the thermodynamic force of diffusion which drives the irreversible process of matter exchange at the macroscopic level. On the other hand, even while the thermodynamic force is a concept of purely statistical origin associated with a collection of particles, in contrast to microscopic forces which have a direct impact on individual particles themselves, it is well known that this force manifests itself as a Newtonian type entry in local conservation laws describing the momentum balance; in fact it pertains to the average (local average) momentum taken over by the particle cloud, a statistical ensemble property quantified in terms of the probability distribution at hand. It is precisely the (negative) gradient of the above potential Q in (2.35) which plays the Newtonian force role in the momentum balance equations. The second analytical expression of interest here involves

$$-\int Q\rho dx = (1/2)\int u^2\rho dx = (1/2)D^2\cdot F_X;\ F_X = \int \frac{(\nabla\rho)^2}{\rho}dx \tag{2.36}$$

where F_X is the Fisher information, encoded in the probability density ρ which quantifies its gradient content (sharpness plus localization/disorder) (note $-\int Q\rho = -\int[(1/2)u^2\rho + D\rho u'] = -\int(1/2)u^2\rho + \int Du\rho' = -(1/2)\int D^2(\rho'/\rho)^2\rho + D^2\int\rho'(\rho'/\rho)$ $= (D^2/2)\int(\rho')^2/\rho = (1/2)\int u^2\rho$). On the other hand the local entropy production inside the system sustaining an irreversible process of diffusion is given via

$$\frac{dS}{dt} = D\cdot\int\frac{(\nabla\rho)^2}{\rho}dx = D\cdot F_X \geq 0 \tag{2.37}$$

This stands for an entropy production rate when the Fick law induced diffusion current (standard Brownian motion case) $j = -D\nabla\rho$, obeying $\partial_t\rho + \nabla j = 0$, enters the scene. Here $S = -\int\rho log(\rho)dx$ plays the role of (time dependent) information entropy in the nonequilibrium statistical mechanics framework for the thermodynamics of irreversible processes. It is clear that a high rate of entropy increase coresponds to a rapid spreading (flattening) of the probability density. This explicitly depends on the sharpness of density gradients. The potential type Q(x,t), the Fisher information F_X, the nonequilibrium measure of entropy production dS/dt, and the information entropy $S(t)$ are thus mutually entangled quantities, each being exclusively determined in terms of ρ and its derivatives.

In the standard statistical mechanics setting the Euler equation gives a prototypical momentum balance equation in the (local) mean

$$(\partial_t + v\cdot\nabla)v = \frac{F}{m} - \frac{\nabla P}{\rho} \tag{2.38}$$

where $F = -\nabla F$ represents normal Newtonian force and P is a pressure term. Q appears in the hydrodynamical formalism of QM via

$$(\partial_t + v\cdot\nabla)v = \frac{1}{m}F - \nabla Q = \frac{1}{m}F + \frac{\hbar^2}{2m^2}\nabla\frac{\Delta\rho^{1/2}}{\rho^{1/2}} \tag{2.39}$$

Another spectacular example pertains to the standard free Brownian motion in the strong friction regime (Smoluchowski diffusion), namely

$$(\partial_t + v\cdot\nabla)v = -2D^2\nabla\frac{\Delta\rho^{1/2}}{\rho^{1/2}} = -\nabla Q \tag{2.40}$$

where $v = -D(\nabla\rho/\rho)$ (formally $D = \hbar/2m$). ■

Remark 2.9. The papers in [39] contain very interesting derivations of Schrödinger equations via diffusion ideas à la Nelson, Markov wave equations, and suitable "applied" forces (e.g. radiative reactive forces). ■

3. Diffusion and Fractals

We go now to Nagasawa [88, 89, 90, 91] to see how diffusion and the SE are really connected (cf. also [3, 10, 23, 57, 93, 97, 111, 119, 117, 120, 121] for related material, some of which is discussed later in detail); for now we simply sketch some formulas for a simple Euclidean metric where (**B1**) $\Delta = \sum(\partial/\partial x^i)^2$. Then $\psi(t,x) = exp[R(t,x) + iS(t,x)]$ satisfies a SE (**B2**) $i\partial_t\psi + (1/2)\Delta\psi + ia(t,x)\cdot\nabla\psi - V(t,x)\psi = 0$ ($\hbar$ and m omitted) if and only if

$$V = -\frac{\partial S}{\partial t} + \frac{1}{2}\Delta R + \frac{1}{2}(\nabla R)^2 - \frac{1}{2}(\nabla S)^2 - a\cdot\nabla S; \tag{3.1}$$

$$0 = \frac{\partial R}{\partial t} + \frac{1}{2}\Delta S + (\nabla S)\cdot(\nabla R) + a\cdot\nabla R$$

in the region (**B3**) $D = \{(s,x) : \psi(s,x) \neq 0\}$. Solutions are often referred to as weak or distributional but we do not belabor this point. From [88, 90] there results

Theorem 3..1. Let $\psi(t,x) = exp[R(t,x) + iS(t,x)]$ be a solution of the SE **(B2)**; then (**B4**) $\phi(t,x) = exp[R(t,x) + S(t,x)]$ and $\hat{\phi} = exp[R(t,x) - S(t,x)]$ are solutions of

$$\frac{\partial\phi}{\partial t} + \frac{1}{2}\Delta\phi + a(t,x)\cdot\nabla\phi + c(t,x,\phi)\phi = 0; \tag{3.2}$$

$$-\frac{\partial\hat{\phi}}{\partial t} + \frac{1}{2}\Delta\hat{\phi} - a(t,x)\cdot\nabla\hat{\phi} + c(t,x,\phi)\hat{\phi} = 0$$

where the creation and annihilation term $c(t,x,\phi)$ is given via

$$c(t,x,\phi) = -V(t,x) - 2\frac{\partial S}{\partial t}(t,x) - (\nabla S)^2(t,x) - 2a\cdot\nabla S(t,x) \tag{3.3}$$

Conversely given $(\phi,\hat{\phi})$ as in **(B4)** satisfying (3.2) it follows that ψ satisfies the SE **(B2)** with V as in (3.3) (note $R = (1/2)log(\hat{\phi}\phi)$ and $S = (1/2)log(\phi/\hat{\phi})$ with $exp(R) = (\hat{\phi}\phi)^{1/2}$).■

We will discuss this later in more detail and give proofs along with probabilistic content (note that the equations (3.2) are not imaginary time SE). From this one can conclude that nonrelativistic QM is diffusion theory in terms of Schrödinger processes (described by $(\phi,\hat{\phi})$ - more details later). Further it is shown that key postulates in Nelson's stochastic mechanics or Zambrini's Euclidean QM (cf. [144]) can both be avoided in connecting the SE to diffusion processes (since they are automatically valid). Look now at Theorem 3.1 for one dimension and write $T = \hbar t$ with $X = (\hbar/\sqrt{m})x$; then the SE **(B2)** becomes (**B5**) $i\hbar\psi_T = -(\hbar^2/2m)\psi_{XX} - iA\psi_X + V\psi$ where $A = a\hbar/\sqrt{m}$. In addition (**B6**) $i\hbar R_T + (\hbar^2/m^2)R_X S_X + (\hbar^2/2m^2)S_{XX} + AR_X = 0$ and (**B7**)$V = -i\hbar S_T + (\hbar^2/2m)R_{XX} + (\hbar^2/2m^2)R_X^2 - (\hbar^2/2m^2)S_X^2 - AS_X$. Hence

Proposition 3.1. Equation **(B2)**, written in the variables (**B8**) $X = (\hbar/\sqrt{m})x$, $T = \hbar t$, with $A = (\sqrt{m}/\hbar)a$ and $V = V(X,T) \sim V(x,t)$ is equivalent to **(B5)**.

Making a change of variables in (3.2) now, as in Proposition 3.1, yields

Corollary 3.1. Equation (3.2), written in the variables of Proposition 3.1, becomes

$$\hbar\phi_T + \frac{\hbar^2}{2m}\phi_{XX} + A\phi_X + \tilde{c}\phi = 0;\ -\hbar\hat{\phi}_T + \frac{\hbar^2}{2m}\hat{\phi}_{XX} - A\hat{\phi}_X + \tilde{c}\hat{\phi} = 0; \tag{3.4}$$

$$\tilde{c} = -\tilde{V}(X,T) - 2\hbar S_T - \frac{\hbar^2}{m} S_X^2 - 2AS_X$$

Thus the diffusion processes pick up factors of $\hbar$ and $\hbar/\sqrt{m}$. ■

Remark 3.1. We extract here from the Appendix to [90] for some remarks on competing points of view regarding diffusion and the the SE. First some work of Fenyes [49] is cited where a Lagrangian is taken as

$$L(t) = \int \left[\frac{\partial S}{\partial t} + \frac{1}{2}(\nabla S)^2 + V + \frac{1}{2}\left(\frac{1}{2}\frac{\nabla \mu}{\mu}\right)^2 \right] \mu dx \quad (3.5)$$

where $\mu_t(x) = exp(2R(t,x))$ denotes the distribution density of a diffusion process and V is a potential function. The term (**B9**) $\Pi(\mu) = (1/2)[(1/2)(\nabla\mu/\mu)]^2$ is called a diffusion pressure and since $(1/2)(\nabla\mu/\mu) \sim \nabla R$ the Lagrangian can be written as

$$L = \int \left[\frac{\partial S}{\partial t} + \frac{1}{2}(\nabla S)^2 + \frac{1}{2}(\nabla R)^2 + V \right] \mu dx \quad (3.6)$$

Applying the variational principle $\delta \int_a^b L(t)dt = 0$ one arrives at

$$\frac{\partial S}{\partial t} + \frac{1}{2}\left[(\nabla(R+S)\right]^2 - (\nabla(R+S)) \cdot \left(\frac{1}{2}\frac{\nabla \mu}{\mu}\right) + \left(\frac{1}{2}\frac{\nabla \mu}{\mu}\right)^2 - \frac{1}{4}\frac{\Delta \mu}{\mu} + V = 0 \quad (3.7)$$

which is called a motion equation of probability densities. From this he shows that the function $\psi = exp(R+iS)$ satisfies the SE (**B10**) $i\partial_t + (1/2)\Delta\psi - V(t,x)\psi = 0$. Indeed putting **(B9)** and the formula (**B11**) $(1/2)(\Delta\mu/\mu) + (1/2)\Delta R + (\nabla R)^2$ into (3.6) one obtains

$$\frac{\partial S}{\partial t} + \frac{1}{2}(\nabla S)^2 - \frac{1}{2}(\nabla R)^2 - \frac{1}{2}\Delta R + V = 0 \quad (3.8)$$

which goes along with the duality relation (**B12**) $R_t + (1/2)\Delta S + \nabla S \cdot \nabla R + b \cdot \nabla R = 0$ where (**B13**) $u = (1/2)(a+\hat{a}) = \nabla R$ and $v = (1/2)(a-\hat{a}) = \nabla S$ as derived in the Nagasawa theory. Hence $\psi = exp(R+iS)$ satisfies the SE by previous calculations. One can see however that the equation (3.6) is not needed since the SE and diffusion equations are equivalent and in fact the equations of motion are the diffusion equations. Moreover it is shown in [90] that (3.6) is an automatic consequence in diffusion theory with $V = -c - 2S_t - (\nabla S)^2$ and therefore it need not be postulated or derived by other means. This is a simple calculation from the theory developed above. ■

Remark 3.2. Nelson's important work in stochastic mechanics [111] produced the SE from diffusion theory but involved a stochastic Newtonian equation which is shown in [90] to be automatically true. Thus Nelson worked in a general context which for our purposes here can be considered in the context of Brownian motions

$$B(t) = \partial_t + (1/2)\Delta + b \cdot \nabla + a \cdot \nabla; \;\; \hat{B}(t) = -\partial_t + (1/2)\Delta - b \cdot \nabla + \hat{a} \cdot \nabla \quad (3.9)$$

and used a mean acceleration (**B14**) $\alpha(t,x) = -(1/2)[B(t)\hat{B}(t)x + \hat{B}(t)B(t)x]$. Assuming the duality relations **(B12)** - **(B13)** he obtains a formula

$$\alpha(t,x) = -\frac{1}{2}[B(t)(-b+\hat{a}) + \hat{B}(b+a)] = b_t + (1/2)\nabla(b)^2 - (b+v) \times curl(b) - \quad (3.10)$$

$-[-v_t + (1/2)\Delta u + (1/2)(\hat{a}\cdot\nabla)a + (1/2)(a\cdot\nabla)\hat{a} - (b\cdot\nabla)v - (v\cdot\nabla)b - v\times curl(b)]$

Then it is shown that the SE can be deduced from the stochastic Newton's equation

$$\alpha(t,x) = -\nabla V + \frac{\partial b}{\partial t} + \frac{1}{2}\nabla(b^2) - (b+v)\times curl(b) \tag{3.11}$$

Nagasawa shows that this serves only to reproduce a known formula for V yielding the SE; he also shows that (3.10) also is an automatic consequence of the duality formulation of diffusion equations above. This equation (3.10) is often called stochastic quantization since it leads to the SE and it is in fact correct with the V specified there. However the SE is more properly considered as following directly from the diffusion equations in duality and is not correctly an equation of motion. There is another discussion of Euclidean QM developed by Zambrini [144]. This involves (**B15**) $\tilde{\alpha}(t,x) = (1/2)[B(t)B(t)x + \hat{B}(t)\hat{B}(t)x]$ (with $(\sigma\sigma^T)^{ij} = \delta^{ij}$). It is postulated that this equals (**B16**) $-\nabla c + b_t + (1/2)\nabla(b)^2 - b + v)\times curl(b)$ which in fact leads to the same equation for V as above with $V = -c - 2S_t - (\nabla S)^2 - 2b\cdot\nabla S$ so there is nothing new. Indeed it is shown in [90] that **(B16)** holds automatically as a simple consequence of time reversal of diffusion processes. ■

3.1. Scale Relativity

There are several excellent and exciting approaches here. The method of Nottale [113, 114, 115] is preeminent (cf. also [119, 120, 121, 122]) and there is also a nice derivation of a nonlinear SE via fractal considerations in [23] (indicated below). The most elaborate and rigorous approach is due to Cresson [33], with elaboration and updating in [2, 34, 35]. We refer here to [14, 13, 26, 33, 34, 113, 114]. There are various derivations of the SE and we follow [114] here (cf. also [115, 135]). The philosophy is discussed in [13, 14, 33, 34, 114] and we just write down equations here. First a bivelocity structure is defined (recall that one is dealing with fractal paths). One defines first

$$\frac{d_+}{dt}y(t) = lim_{\Delta t\to 0_+}\left\langle \frac{y(t+\Delta t) - y(t)}{\Delta t}\right\rangle; \tag{3.12}$$

$$\frac{d_-}{dt}y(t) = lim_{\Delta t\to 0_+}\left\langle \frac{y(t) - y(t-\Delta t)}{\Delta t}\right\rangle$$

Applied to the position vector x this yields forward and backward mean velocities, namely (**B17**) $(d_+/dt)x(t) = b_+$ and $(d_-/dt)x(t) = b_-$. Here these velocities are defined as the average at a point q and time t of the respective velocities of the outgoing and incoming fractal trajectories; in stochastic QM this corresponds to an average on the quantum state. The position vector $x(t)$ is thus "assimilated" to a stochastic process which satisfies respectively after $(dt > 0)$ and before $(dt < 0)$ the instant t a relation (**B18**) $dx(t) = b_+[x(t)]dt + d\xi_+(t) = b_-[x(t)]dt + d\xi_-(t)$ where $\xi(t)$ is a Wiener process (cf. [111]). It is in the description of ξ that the $D = 2$ fractal character of trajectories is inserted; indeed that ξ is a Wiener process means that the $d\xi$'s are assumed to be Gaussian with mean 0, mutually independent, and such that

$$< d\xi_{+i}(t)d\xi_{+j}(t) >= 2\mathcal{D}\delta_{ij}dt;\ < d\xi_{-i}(t)d\xi_{-j}(t) >= -2\mathcal{D}\delta_{ij}dt \tag{3.13}$$

where $< >$ denotes averaging and $\mathcal{D}$ is the diffusion coefficient. Nelson's postulate (cf. [111]) is that $\mathcal{D} = \hbar/2m$ and this has considerable justification (cf. [114]). Note also that (3.13) is indeed a consequence of fractal (Hausdorff) dimension 2 of trajectories follows from $< d\xi^2 > /dt^2 = dt^{-1}$, i.e. precisely Feynman's result $< v^2 >^{1/2} \sim \delta t^{-1/2}$ (the discussion here in [114] is unclear however - cf. [1]). Note also that Brownian motion (used in Nelson's postulate) is known to be of fractal (Hausdorff) dimension 2. Note also that any value of $\mathcal{D}$ may lead to QM and for $\mathcal{D} \to 0$ the theory becomes equivalent to the Bohm theory. Now expand any function $f(x,t)$ in a Taylor series up to order 2, take averages, and use properties of the Wiener process ξ to get

$$\frac{d_+ f}{dt} = (\partial_t + b_+ \cdot \nabla + \mathcal{D}\Delta)f; \;\; \frac{d_- f}{dt} = (\partial_t + b_- \cdot \nabla - \mathcal{D}\Delta)f \tag{3.14}$$

Let $\rho(x,t)$ be the probability density of $x(t)$; it is known that for any Markov (hence Wiener) process one has $(\mathbf{B19})$ $\partial_t\rho + div(\rho b_+) = \mathcal{D}\Delta\rho$ (forward equation) and $(\mathbf{B20})$ $\partial_t\rho + div(\rho b_-) = -\mathcal{D}\Delta\rho$ (backward equation). These are called Fokker-Planck equations and one defines two new average velocities $(\mathbf{B21})$ $V = (1/2)[b_+ + b_-]$ and $U = (1/2)[b_+ - b_-]$. Consequently adding and subtracting one obtains $(\mathbf{B22})$ $\rho_t + div(\rho V) = 0$ (continuity equation) and $(\mathbf{B23})$ $div(\rho U) - \mathcal{D}\Delta\rho = 0$ which is equivalent to $(\mathbf{B24})$ $div[\rho(U - \mathcal{D}\nabla log(\rho))] = 0$. One can show, using (3.14) that the term in square brackets in **(B24)** is zero leading to $(\mathbf{B25})$ $U = \mathcal{D}\nabla log(\rho)$. Now place oneself in the (U,V) plane and write $(\mathbf{B26})$ $\mathcal{V} = V - iU$. Then write $(\mathbf{B27})$ $(d_{\mathcal{V}}/dt) = (1/2)(d_+ + d_-)/dt$ and $(d_{\mathcal{U}}/dt) = (1/2)(d_+ - d_-)/dt$. Combining the equations in (3.14) one defines $(\mathbf{B28})$ $(d_{\mathcal{V}}/dt) = \partial_t + V \cdot \nabla$ and $(d_{\mathcal{U}}/dt) = \mathcal{D}\Delta + U \cdot \nabla$; then define a complex operator $(\mathbf{B29})$ $(d'/dt) = (d_{\mathcal{V}}/dt) - i(d_{\mathcal{U}}/dt)$ which becomes

$$\frac{d'}{dt} = \left(\frac{\partial}{\partial t} - i\mathcal{D}\Delta\right) + \mathcal{V} \cdot \nabla \tag{3.15}$$

One now postulates that the passage from classical mechanics to a new nondifferentiable process considered here can be implemented by the unique prescription of replacing the standard d/dt by d'/dt. Thus consider $(\mathbf{B30})$ $\mathcal{S} = \left\langle \int_{t_1}^{t_2} \mathcal{L}(x, \mathcal{V}, t)dt \right\rangle$ yielding by least action $(\mathbf{B31})$ $(d'/dt)(\partial\mathcal{L}/\partial\mathcal{V}_i) = \partial\mathcal{L}/\partial x_i$. Define then $\mathcal{P}_i = \partial\mathcal{L}/\partial\mathcal{V}_i$ leading to $(\mathbf{B32})$ $\mathcal{P} = \nabla\mathcal{S}$ (note this is $\mathcal{S}$ and not S). Now for Newtonian mechanics write $(\mathbf{B33})$ $L(x,v,t) = (1/2)mv^2 - \mathbf{U}$ which becomes $\mathcal{L}(x,\mathcal{V},t) = (1/2)m\mathcal{V}^2 - \mathfrak{U}$ leading to $(\mathbf{B34})$ $-\nabla\mathfrak{U} = m(d'/dt)\mathcal{V}$. One separates real and imaginary parts of the complex acceleration $\gamma = (d'\mathcal{V}/dt$ to get

$$d'\mathcal{V} = (d_{\mathcal{V}} - id_{\mathcal{U}})(V - iU) = (d_{\mathcal{V}}V - d_{\mathcal{U}}U) - i(d_{\mathcal{U}}V + d_{\mathcal{V}}U) \tag{3.16}$$

The force $F = -\nabla\mathfrak{U}$ is real so the imaginary part of the complex acceleration vanishes; hence

$$\frac{d_{\mathcal{U}}}{dt}V + \frac{d_{\mathcal{V}}}{dt}U = \frac{\partial U}{\partial t} + U \cdot \nabla V + V \cdot \nabla U + \mathcal{D}\Delta V = 0 \tag{3.17}$$

from which $\partial U/\partial t$ may be obtained. Differentiating the expression $U = \mathcal{D}\nabla log(\rho)$ and using the continuity equation yields another expression $(\mathbf{B35})$ $(\partial U/\partial t) = -\mathcal{D}\nabla(divV) - \nabla(V \cdot U)$. Comparison of these relations yields $\nabla(divV) = \Delta V - U \wedge curlV$ where the

$curlU$ term vanishes since U is a gradient. However in the Newtonian case $\mathcal{P} = m\mathcal{V}$ so **(B32)** implies that $\mathcal{V}$ is a gradient and hence a generalization of the classical action S can be defined via $(\mathbf{B36})$ $V = 2\mathcal{D}\nabla S$ (note then $\nabla(divV) = \Delta V$ and $curlV = 0$). Combining this with the expression for U one obtains $(\mathbf{B37})$ $\mathcal{S} = log(\rho^{1/2}) + iS$. One notes that this is compatible with [111] for example. The way to the SE is now short; set $(\mathbf{B38})$ $\psi = \sqrt{\rho}exp(iS) = exp(i\mathcal{S})$ with $(\mathbf{B39})$ $\mathcal{V} = -2i\mathcal{D}\nabla(log\psi)$ (note $U = \mathcal{D}\nabla log(\rho)$, $V = 2\mathcal{D}\nabla S$, $\mathcal{V} = -2i\mathcal{D}\nabla log\psi = -i\mathcal{D}\nabla log(\rho) + 2\mathcal{D}\nabla S = V - iU$); thus for $\mathcal{P} = m\mathcal{V}$ the relation $(\mathbf{B40})$ $\mathcal{P} \sim -i\hbar\nabla$ or $\mathcal{P}\psi = -i\hbar\nabla\psi$ has a natural interpretation. Putting ψ in **(B34)**, which generalizes Newton's law to fractal space the equation of motion takes the form $(\mathbf{B41})$ $\nabla\mathfrak{U} = 2i\mathcal{D}m(d'/dt)(\nabla log(\psi))$. Noting that d' and ∇ do not commute one replaces d'/dt by (3.15) to obtain

$$\nabla\mathfrak{U} = 2i\mathcal{D}m\left[\partial_t\nabla log(\psi) - i\mathcal{D}\Delta(\nabla log(\psi)) - 2i\mathcal{D}(\nabla log(\psi)\cdot\nabla)(\nabla log(\psi)\right] \quad (3.18)$$

This expression can be simplified via

$$\nabla\Delta = \Delta\nabla;\ (\nabla f\cdot\nabla)(\nabla f) = (1/2)\nabla(\nabla f)^2;\ \frac{\Delta f}{f} = \Delta log(f) + (\nabla log(f))^2 \quad (3.19)$$

This implies

$$\frac{1}{2}\Delta(\nabla log(\psi)) + (\nabla log(\psi)\cdot\nabla)(\nabla log(\psi)) = \frac{1}{2}\nabla\frac{\Delta\psi}{\psi} \quad (3.20)$$

Integrating this equation yields $(\mathbf{B42})$ $\mathcal{D}^2\Delta\psi + i\mathcal{D}\partial_t\psi - (\mathfrak{U}/2m)\psi = 0$ up to an arbitrary phase factor $\alpha(t)$ which can be set equal to 0 by a suitable choice of phase S. Replacing $\mathcal{D}$ by $\hbar/2m$ one arrives at the SE $(\mathbf{B43})$ $i\hbar\psi_t = -(\hbar^2/2m)\Delta\psi + \mathfrak{U}\psi$. This suggests an interpretation of QM as mechanics in a nondifferentiable (fractal) space.

Remark 3.3. Some of the relevant equations for dimension one are collected together in Section 6. We note that it is the presence of $\pm$ derivatives that makes possible the introduction of a complex plane to describe velocities and hence QM; one can think of this as the motivation for a complex valued wave function and the nature of the SE. ∎

We go now to [23] and will sketch some of the material. Here one extends ideas of Nottale and Ord in order to derive a nonlinear Schrödinger equation (NLSE). Using the hydrodynamic model in [124] one added a hydrostatic pressure term to the Euler-Lagrange equations and another possibility is to add instead a kinematic pressure term. The hydrostatic pressure is based on an Euler equation $-\nabla p = \rho g$ where ρ is density and g the gravitational acceleration (note this gives $p = \rho g x$ in 1-D). In [124] one took $\rho = \psi^*\psi$, b a mass-energy parameter, and $p = \rho$; then the hydrostatic potential is (for $\rho_0 = 1$)

$$b\int g(x)\cdot dr = -b\int\frac{\nabla p}{\rho}\cdot dr = -blog(\rho/\rho_0) = -blog(\psi^*\psi) \quad (3.21)$$

Here $-blog(\psi^*\psi)$ has energy units and explains the nonlinear term of [9] which involved

$$i\hbar\frac{\partial\psi}{\partial t} = -\frac{\hbar^2}{2m}\nabla^2\psi + U\psi - b[log(\psi^*\psi)]\psi \quad (3.22)$$

A derivation of this equation from the Nelson stochastic QM was given by Lemos (cf. [79]). There are however some problems since this equation does not obey the homogeneity condition saying that the state $\lambda|\psi>$ is equivalent to $|\psi>$; however (3.22) is not invariant under $\psi \to \lambda\psi$. Further, plane wave solutions to (3.22) do not seem to have a physical interpretion due to extraneous dispersion relations. Finally one would like to have a SE in terms of ψ alone. Note that another NLSE could be obtained by adding kinetic pressure terms $(1/2)\rho v^2$ and taking $\rho = a\psi^*\psi$ where $v = p/m$. Now using the relations from HJ theory (**B44**) $(\psi/\psi^*) = exp[2i\mathfrak{S}(x)/\hbar]$ and $p = \nabla\mathfrak{S}(x) = mv$ one can write (**B45**) $v = -i(\hbar/2m)\nabla log(\psi/\psi^*)$ so that the energy density becomes (**B46**) $(1/2)\rho|v|^2 = (a\hbar^2/8m^2)\psi\psi^*\nabla log(\psi/\psi^*)\cdot\nabla log(\psi^*/\psi)$. This leads to a corresponding nonlinear potential associated with the kinematical pressure via (**B47**) $(a\hbar^2/8m^2)\nabla log(\psi/\psi^*)\cdot\nabla log(\psi^*/\psi)$. Hence a candidate NLSE is

$$i\hbar\partial_t = -\frac{\hbar^2}{2m}\nabla^2\psi + U\psi - b[log(\psi^*\psi)]\psi + \frac{a\hbar^2}{8m^2}\left(\nabla log\frac{\psi}{\psi^*}\cdot\nabla log\frac{\psi^*}{\psi}\right) \quad (3.23)$$

(apparently this equation has not yet been derived in the literature). Here the Hamiltonian is Hermitian and $a \neq b$ are both mass-energy parameters to be determined experimentally. The new term can also be written in the form (**B48**) $\nabla log(\psi/\psi^*)\cdot\nabla log(\psi^*/\psi) = -[\nabla log(\psi/\psi^*)]^2$. The goal now is to derive a NLSE directly from fractal space time dynamics for a particle undergoing Brownian motion. This does not require a quantum potential, a hydrodynamic model, or any pressure terms as above.

Remark 3.4. One should make some comments about the kinematic pressure terms (**B49**) $(1/2)\rho v^2 \iff (\hbar^2/2m)(a/m)|\nabla log(\psi)|^2$ versus hydrostatic pressure terms of the form (**B50**) $\int(\nabla p/\rho) \iff -blog(\psi^*\psi)$. The hydrostatic term breaks homogeneity whereas the kinematic pressure term preserves homogeneity (scaling with a λ factor). The hydrostatic pressure term is also not compatible with the motion kinematics of a particle executing a fractal Brownian motion. The fractal formulation will enable one to relate the parameters a, b to $\hbar$. ∎

Following Nottale nondifferentiability implies a loss of causality and one is thinking of Feynmann paths with $< v^2 > \propto (dx/dt)^2 \propto dt^{2[(1/D)-1)}$ with $D = 2$. Now a fractal function $f(x,\epsilon)$ could have a derivative $\partial f/\partial\epsilon$ and renormalization group arguments lead to (**B51**) $(\partial f(x,\epsilon)/\partial log\epsilon) = a(x) + bf(x,\epsilon)$ (cf. [114]). This can be integrated to give (**B52**) $f(x,\epsilon) = f_0(x)[1-\zeta(x)(\lambda/\epsilon)^{-b}]$. Here $\lambda^{-b}\zeta(x)$ is an integration constant and $f_0(x) = -a(x)/b$. This says that any fractal function can be approximated by the sum of two terms, one independent of the resolution and the other resolution dependent; one expects $\zeta(x)$ to be a flucuating function with zero mean. Provided $a \neq 0$ and $b < 0$ one has two interesting cases (i) $\epsilon << \lambda$ with $f(x,\epsilon) \sim f_0(x)(\lambda/\epsilon)^{-b}$ and (ii) $\epsilon >> \lambda$ with f independent of scale. Here λ is the deBroglie wavelength. Now one writes

$$r(t+dt,dt) - r(t,dt) = b_+(r,t)dt + \xi_+(t,dt)\left(\frac{dt}{\tau_0}\right)^{\beta}; \quad (3.24)$$

$$r(t,dt) - r(t-dt,dt) - b_-(r,t)dt + \xi_-(t,dt)\left(\frac{dt}{\tau_0}\right)^{\beta}$$

where $\beta = 1/D$ and $b_\pm$ are average forward and backward velocities. This leads to $(\mathbf{B53})$ $v_\pm(r,t,dt) = b_\pm(r,t) + \xi_\pm(t,dt)(dt/\tau_0)^{\beta-1}$. In the quantum case $D = 2$ one has $\beta = 1/2$ so $dt^{\beta-1}$ is a divergent quantity (so nondifferentiability ensues). Following [79, 114, 111] one defines

$$\frac{d_\pm r(t)}{dt} = lim_{\Delta t \to \pm 0}\left\langle \frac{r(t+\Delta t) - r(t)}{\Delta t}\right\rangle \tag{3.25}$$

from which $(\mathbf{B54})$ $d_\pm r(t)/dt = b_\pm$. Now following Nottale one writes

$$\frac{\delta}{dt} = \frac{1}{2}\left(\frac{d_+}{dt} + \frac{d_-}{dt}\right) - \frac{i}{2}\left(\frac{d_+}{dt} - \frac{d_-}{dt}\right) \tag{3.26}$$

which leads to $(\mathbf{B55})$ $(\delta/dt) = (\partial/\partial t) + v\cdot\nabla - iD\nabla^2$. Here in principle D is a real valued diffusion constant to be related to $\hbar$. (A symbol D for the fractal dimension is no longer needed here (?) - e.g. $D = 2$ with $(\mathbf{B56})$ $< d\xi_{\pm i}d\xi_{\pm j} >= \pm 2D\delta_{ij}dt$.) Now for the complex time dependent wave function we take $\psi = exp[i\mathfrak{S}/2mD]$ with $p = \nabla\mathfrak{S}$ so that $(\mathbf{B57})$ $v = -2iD\nabla log(\psi)$. The SE is obtained from the Newton equation ($F = ma$) via $(\mathbf{B58})$ $-\nabla U = m(\delta/dt)v = -2imD(\delta/dt)\nabla log(\psi)$. Inserting **(B55)** gives

$$-\nabla U = -2im[D\partial_t\nabla log(\psi)] - 2D\nabla\left(D\frac{\nabla^2\psi}{\psi}\right) \tag{3.27}$$

(see [114] for identities involving ∇). Integrating (3.27) yields $(\mathbf{B59})$ $D^2\nabla^2\psi + iD\partial_t\psi - (U/2m)\psi = 0$ up to an arbitrary phase factor which may be set equal to zero. Now replacing D by $\hbar/2m$ one gets the SE $(\mathbf{B60})$ $i\hbar\partial_t\psi + (\hbar^2/2m)\nabla^2\psi = U\psi$. Here the Hamiltonian is Hermitian, the equation is linear, and the equation is homogeneous of degree 1 under the substitution $\psi \to \lambda\psi$.

Next one generalizes this by relaxing the assumption that the diffusion coefficient is real. Some comments on complex energies are needed - in particular constraints are often needed (cf. [129]). However complex energies are not alien in ordinary QM (cf. [23] for references). Now the imaginary part of the linear SE yields the continuity equation $\partial_t\rho + \nabla\cdot(\rho v) = 0$ and with a complex potential the imaginary part of the potential will act as a source term in the continuity equation. Instead of $(\mathbf{B61})$ $< d\zeta_\pm d\zeta_\pm >= \pm 2Ddt$ with D and $2mD = \hbar$ real one sets $(\mathbf{B62})$ $< d\zeta_\pm d\zeta_\pm >= \pm(D + D^*)dt$ with D and $2mD = \hbar = \alpha + i\beta$ complex. The complex time derivative operator becomes $(\mathbf{B63})$ $(\delta/dt) = \partial_t + v\cdot\nabla - (i/2)(D + D^*)\nabla^2$. Writing again $(\mathbf{B64})$ $\psi = exp[i\mathfrak{S}/2mD] = exp(i\mathfrak{S}/\hbar)$ one obtains $(\mathbf{B65})$ $v = -2iD\nabla log(\psi)$. The NLSE is then obtained (via the Newton law) as $(\mathbf{B66})$ $-\nabla U = m(\delta/dt)v = -2imD(\delta/dt)\nabla log(\psi)$. Inserting **(B63)** one gets

$$\nabla U = 2im\left[D\partial_t\nabla log(\psi) - 2iD^2(\nabla log(\psi)\cdot\nabla)(\nabla log(\psi) - \frac{i}{2}(D + D^*)D\nabla^2(\nabla log(\psi)\right] \tag{3.28}$$

Now using the identities (i) $\nabla\nabla^2 = \nabla^2\nabla$, (ii) $2(\nabla log(\psi)\cdot\nabla)(\nabla log(\psi) = \nabla(\nabla log(\psi))^2$ and (iii) $\nabla^2 log(\psi) = \nabla^2\psi/\psi - (\nabla log(\psi))^2$ leads to a NLSE with nonlinear (kinematic pressure) potential, namely

$$i\hbar\partial_t\psi = -\frac{\hbar^2}{2m}\frac{\alpha}{\hbar}\nabla^2\psi + U\psi - i\frac{\hbar^2}{2m}\frac{\beta}{\hbar}(\nabla log(\psi))^2\psi \tag{3.29}$$

Note the crucial minus sign in front of the kinematic pressure term and also that $\hbar = \alpha + i\beta = 2mD$ is complex. When $\beta = 0$ one recovers the linear SE. The nonlinear potential is complex and one defines **(B67)** $W = -(\hbar^2/2m)(\beta/\hbar)(\nabla log(\psi))^2$ with U the ordinary potential; then the NLSE is **(B68)** $i\hbar\partial_t\psi = [-(\hbar^2/2m)(\alpha/\hbar)\nabla^2 + U + iW]\psi$. This is the fundamental result of [23]; it has the form of an ordinary SE with complex potential $U + iW$ and complex $\hbar$. The Hamiltonian is no longer Hermitian and the potential itself depends on ψ. Nevertheless one can have meaningful physical solutions with real valued energies and momenta; the homogeneity breaking hydrostatic pressure term $-b(log(\psi^*\psi)\psi$ is not present (it would be meaningless) and the NLSE is invariant under $\psi \to \lambda\psi$.

Remark 3.5. One could ask why not simply propose as a valid NLSE an equation

$$i\hbar\partial_t\psi = -\frac{\hbar^2}{2m}\nabla^2\psi + U\psi + \frac{\hbar^2}{2m}\frac{a}{m}|\nabla log(\psi)|^2\psi \tag{3.30}$$

Here one has a real Hamiltonian satisfying the homogeneity condition and the equation admits soliton solutions of the form **(B69)** $\psi = CA(x - vt)exp[i(kx - \omega t)]$ where $A(x - vt)$ is to be determined by solving the NLSE. The problem here is that the equation suffers from an extraneous dispersion relation. Thus putting in the plane wave solution $\psi \sim exp[-i(Et - px)]$ one gets an extraneous EM relation (after setting $U = 0$), namely **(B70)** $E = (p^2/2m)[1 + (a/m)]$ instead of the usual $E = p^2/2m$ and hence $E_{QM} \neq E_{FT}$ where FT means field theory. ■

Remark 3.6. It has been known since e.g. [129] that the expression for the energy functional in nonlinear QM does not coincide with the QM energy functional, nor is it unique. To see this write down the NLSE of [9] in the form **(B71)** $i\hbar\partial_t\psi = \partial H(\psi, \psi^*)/\partial\psi^*$ where the real Hamiltonian density is

$$H(\psi, \psi^*) = -\frac{\hbar^2}{2m}\psi^*\nabla^2\psi + U\psi^*\psi - b\psi^* log(\psi^*\psi)\psi + b\psi^*\psi \tag{3.31}$$

Then using $E_{FT} = \int H d^3r$ we see it is different from $< \hat{H} >_{QM}$ and in fact $E_{FT} - E_{QM} = \int b\psi^*\psi d^3r = b$. This problem does not occur in the fractal based NLSE since it is written entirely in terms of ψ. ■

Remark 3.7. In the fractal based NLSE there is no discrepancy between the QM energy functional and the FT energy functional. Both are given by

$$N^{NLSE}_{fractal} = -\frac{\hbar^2}{2m}\frac{\alpha}{\hbar}\psi^*\nabla^2\psi + U\psi^*\psi - i\frac{\hbar^2}{2m}\frac{\beta}{\hbar}\psi^*(\nabla log(\psi)^2\psi \tag{3.32}$$

The NLSE is unambiguously given by **(B71)** and $H(\psi, \psi^*)$ is homogeneous of degree 1 in λ. Such equations admit plane wave solutions with dispersion relation $E = p^2/2m$; indeed, inserting the plane wave solution into the fractal based NLSE one gets (after setting $U = 0$)

$$E = \frac{\hbar^2}{2m}\frac{\alpha}{\hbar}\frac{p^2}{2m} + i\frac{\beta}{\hbar}\frac{p^2}{2m} = \frac{p^2}{2m}\frac{\alpha + i\beta}{\hbar} = \frac{p^2}{2m} \tag{3.33}$$

since $\hbar = \alpha + i\beta$. The remarkable feature of the fractal approach versus all other NLSE considered sofar is that the QM energy functional is precisely the FT one. The complex diffusion constant represents a truly new physical phenomenon insofar as a small imaginary correction to the Planck constant is the hallmark of nonlinearity in QM (see [23] for more on this). ■

4. Remarks on a Fractal Spacetime

There have been a number of articles and books involving fractal methods in spacetime or fractal spacetime itself with impetus coming from quantum physics and relativity. We refer here especially to [1, 14, 13, 24, 58, 94, 95, 96, 97, 98, 99, 100, 101, 102, 103, 105, 106] for background to this paper. Many related papers are omitted here and we refer in particular to the journal Chaos, Solitons, and Fractals CSF) for further information. For information on fractals and stochastic processes we refer for example to [4, 5, 27, 28, 29, 46, 59, 74, 75, 80, 85, 110, 117, 125, 127, 134, 138, 139, 143]. We discuss here a few background ideas and constructions in order to indicate the ingredients for El Naschie's Cantorian spacetime $\mathfrak{E}^{\infty}$, whose exact nature is elusive. Suitable references are given but there are many more papers in the journal CSF by El Naschie (and others) based on these fundamental ideas and these are either important in a revolutionary sense or a fascinating refined form of science fiction. In what appears at times to be pure numerology one manages to (rather hastily) produce amazingly close numerical approximations to virtually all the fundamental constants of physics (including string theory). The key concepts revolve around the famous golden ratio $(\sqrt{5}-1)/2$ and a strange Cantorian space $\mathfrak{E}^{\infty}$ which we try to describe below. It is very tempting to want all of these (heuristic) results to be true and the approach seems close enough and universal enough to compel one to think something very important must be involved. Moreover such scope and accuracy cannot be ignored so we try to examine some of the constructions in a didactic manner in order to possibly generate some understanding.

4.1. Comments on Cantor Sets

Example 4.1. In the paper [85] one discusses random recursive constructions leading to Cantor sets, etc. Associated with each such construction is a universal number α such that almost surely the random object has Hausdorff dimension α (we assume that ideas of Hausdorff and Minkowski-Bouligand (MB) or upper box dimension are known - cf. [5, 14, 46, 80]). One construction of a Cantor set goes as follows. Choose x from $[0, 1]$ according to the uniform distribution and then choose y from $[x, 1]$ according to the uniform distribution on $[x, 1]$. Set $J_0 = [0, x]$ and $J_1 = [y, 1]$ and recall the standard $1/3$ construction for Cantor sets. Continue this procedure by rescaling to each of the intervals already obtained. With probability one one then obtains a Cantor set S_c^0 with Hausdorff dimension **(C1)** $\alpha = \phi = (\sqrt{5}-1)/2 \sim .618$. Note that this is just a particular random Cantor set; there are others with different Hausdorff dimensions (there seems to be some - possibly harmless - confusion on this point in the El Naschie papers). However the golden ratio ϕ is a very interesting number whose importance rivals that of π or e. In particular (cf. [4]) ϕ is the hardest number to approximate by rational numbers and could be called the most irrational number. This is because its continued fraction represention involves all $1's$. ■

Example 4. 2. From [94] the Hausdorff (H) dimension of a traditional triadic Cantor set is $d_c^{(0)} = log(2)/log(3)$. To determine the equivalent to a triadic Cantor set in 2 dimensions one looks for a set which is triadic Cantorian in all directions. The analogue of an area $A = 1 \times 1$ is a quasi-area $A_c = d_c^{(0)} \times d_c^{(0)}$ and to normalize A_c one uses $\rho_2 = (A/A_c)_2 = 1/(d_c^{(0)})^2$ (for n-dimensions (**C2**) $\rho_n = 1/(d_c^{(0)})^{n-1}$). Then the nth Cantor like H dimension $d_c^{(n)}$ will have the form (**C3**) $d_c^{(n)} = \rho_n d_c^{(0)} = 1/(d_c^{(0)})^{n-1}$. Note also that the H dimension of a Sierpinski gasket is (**C4**) $d_c^{(n+1)}$ $/d_c^{(n)} = 1/d_c^{(0)} = log(3)/log(2)$ and in any event the straight-forward interpretation of $d_c^{(2)} = log(3)/log(2)$ is a scaling of $d_c^{(0)} = log(2)/log(3)$ proportional to the ratio of areas $(A/A_c)_2$. One notes that (**C5**) $d_c^{(4)} = 1/(d_c^{(0)})^3 = (log(3)/log(2))^3 \simeq 3.997 \sim 4$ so the 4-dimensional Cantor set is essentially "space filling".

Another derivation goes as follows. Define probability quotients $\Omega = dim(subset)$ $/dim(set)$. For a triadic Cantor set in 1-D (**C6**) $\Omega^{(1)} = d_c^{(0)}/d_c^{(1)} = d_c^{(0)}$ $(d_c^{(1)} = 1)$. To lift the Cantor set to n-dimensions look at the multiplicative probability law (**C7**) $\Omega^{(n)} = (\Omega^{(1)})^n = (d_c^{(0)})^n$. However since $\Omega^{(1)} = d_c^{(0)}/d_c^{(n)}$ we get (**C8**) $d_c^{(0)}/d_c^{(n)} = (d_c^{(0)})^n \Rightarrow d_c^{(n)} = 1/(d_c^{(0)})^{n-1}$. Since $\Omega^{(n-1)}$ is the probability of finding a Cantor point (Cantorian) one can think of the H dimension $d_c^{(n)} = 1/\Omega^{(n-1)}$ as a measure of ignorance. One notes here also that for $d_c^{(0)} = \phi$ (the Cantor set $S_c^{(0)}$ of Example 2.1) one has $d_c^{(4)} = 1/\phi^3 = 4 + \phi^3 \simeq 4.236$ which is surely space filling. ■

Based on these ideas one proves in [95, 96, 98] a number of theorems and we sketch some of this here. One picks a "backbone" Cantor set with H dimension $d_c^{(0)}$ (the choice of $\phi = d_c^{(0)}$ will turn out to be optimal for many arguments). Then one imagines a Cantorian spacetime $\mathfrak{E}^\infty$ built up of an infinite number of spaces of dimension $d_c^{(n)}$ $(-\infty \leq n < \infty)$. The exact form of embedding etc. here is not specified so one imagines e.g. $\mathfrak{E}^\infty = \cup \mathfrak{E}^{(n)}$ (with unions and intersections) in some amorphous sense. There are some connections of this to vonNeumann's continuous geometries indicated in [100]. In this connection we remark that only $\mathfrak{E}^{(-\infty)}$ is the completely empty set ($\mathfrak{E}^{(-1)}$ is not empty). First we note that $\phi^2 + \phi - 1 = 0$ leading to (**C9**) $1 + \phi = 1/\phi,\ \phi^3 = (2+\phi)/\phi,\ (1+\phi)/(1-\phi) = 1/\phi(1-\phi) = 4 + \phi^3 = 1/\phi^3$ (a very interesting number indeed). Then one asserts that

Theorem 4.1. Let $(\Omega^{(1)})^n$ be a geometrical measure in n-dimensional space of a multiplicative point set process and $\Omega^{(1)}$ be the Hausdorff dimension of the backbone (generating) set $d_c^{(0)}$. Then $< d >= 1/d_c^{(0)}(1 - d_c^{(0)})$ (called curiously an average Hausdorff dimension) will be exactly equal to the average space dimension $< n >= (1 + d_c^{(0)})(1 - d_c^{(0)})$ and equivalent to a 4-dimensional Cantor set with H-dimension $d_c^{(4)} = 1/(d_c^{(0)})^3$ if and only if $d_c^{(0)} = \phi$.

To see this take $\Omega^{(n)} = (\Omega^{(1)})^n$ again and consider the total probability of the additive set described by the $\Omega^{(n)}$, namely (**C10**) $Z_0 = \sum_0^\infty (\Omega^{(1)})^n = 1/(1 - \Omega^{(1)})$. It is conceptually easier here to regard this as a sum of weighted dimensions (since $d_c^{(n)} = 1/(d_c^{(0)})^{n-1}$) and consider $w_n = n(d_c^{(0)})^n$. Then the expectation of n becomes (note $d_c^{(n)} \sim 1/(d_c^{(0)})^{n-1} \sim 1/\Omega^{(n-1)}$ so $n(d_c^{(0)})^{n-1} \sim n/d_c^{(n)}$)

$$E(n) = \frac{\sum_1^\infty n^2 (d_c^{(0)})^{n-1}}{\sum_1^\infty n(d_c^{(0)})^{n-1}} = \tilde{<} n >= \frac{1+d_c^{(0)}}{1-d_c^{(0)}} \tag{4.1}$$

Another average here is defined via (blackbody gamma distribution)

$$< n >= \frac{\int_0^\infty n^2 (\Omega^{(1)})^n dn}{\int_0^\infty n(\Omega^{(1)})^n dn} = \frac{-2}{log(\Omega^{(1)})} \tag{4.2}$$

which corresponds to $\tilde{<} n >$ after expanding the logarithm and omitting higher order terms. However $\tilde{<} n >$ seems to be the more valid calculation here. Similarly one defines (somewhat ambiguously) an expected value for $d_c^{(n)}$ via

$$< d >= \frac{\sum_1^\infty n(d_c^{(0)})^{n-1}}{\sum_1^\infty (d_c^{(0)})^n} = \frac{1}{d_c^{(0)}(1-d_c^{(0)})} \tag{4.3}$$

This is contrived of course (and cannot represent $E(d_c^{(n)})$ since one is computing reciprocals $\sum(n/d_c^{(n)})$ but we could think of computing an expected ignorance and identifying this with the reciprocal of dimension. Thus the label $< d >$ does not seem to represent an expected dimension but if we accept it as a symbol then for $d_c^{(0)} = \phi$ one has from **(C9)**

$$\tilde{<} n >= \frac{1+\phi}{1-\phi} =< d >= \frac{1}{\phi(1-\phi)} = d_c^{(4)} = 4 + \phi^3 = \frac{1}{\phi^3} \sim 4.236 \tag{4.4}$$

Remark 4.1. We note that the normalized probability **(C11)** $N = \Omega^{(1)}/Z_0 = \Omega^{(1)}(1-\Omega^{(1)}) = 1/ < d >$ for any $d_c^{(0)}$. Further if $< d >= 4 = 1/d_c^{(0)}(1-d_c^{(0)})$ one has $d_c^{(0)} = 1/2$ while $\tilde{<} n >= 3 < 4 =< d >$. One sees also that $d_c^{(0)} = 1/2$ is the minimum (where $d < d > /d(d_c^{(0)}) = 0$). ■

Remark 4.2. The results of Theorem 4.1 should really be phrased in terms of $\mathfrak{E}^\infty$ (cf. [101]). thus ($H \sim$ Hausdorff dimension and $T \sim$ topological dimension)

$$dim_H \mathfrak{E}^{(n)} = d_c^{(n)} = \frac{1}{(d_c^{(0)})^{n-1}}; \tag{4.5}$$

$$< d >= \frac{1}{d_c^{(0)}(1-d_c^{(0)})}; \tilde{<} dim_T \mathfrak{E}^\infty >= \frac{1+d_c^{(0)}}{1-d_c^{(0)}} = \tilde{<} n >$$

In any event $\mathfrak{E}^\infty$ is formally infinite dimensional but effectively it is $4\pm$ dimensional with an infinite number of internal dimensions. We emphasize that $\mathfrak{E}^\infty$ appears to be constructed from a fixed backbone Cantor set with H dimension $1/2 \leq d_c^{(0)} < 1$; thus each such $d_c^{(0)}$ generates an $\mathfrak{E}^\infty$ space. Note that in [101] $\mathfrak{E}^\infty$ is looked upon as a transfinite discretum (?) underpinning the continuum. ■

Remark 4.3. An interesting argument from [100] goes as follows. Thinking of $d_c^{(0)}$ as a geometrical probability one could say that the spatial (3-dimensional) probability of finding a Cantorian "point" in $\mathfrak{E}^{\infty}$ must be given by the intersection probability $(\mathbf{C12})\ P = (d_c^{(0)})^3$ where $3 \sim 3$ topological spatial dimension. P could then be regarded as a Hurst exponent (cf. [1, 114, 143]) and the Hausdorff dimension of the fractal path of a Cantorian would be $(\mathbf{C13})\ d_{path} = 1/H = 1/P = 1/(d_c^{(0)})^3$. Given $d_c^{(0)} = \phi$ this means $d_{path} = 4+\phi^3 \sim 4^+$ so a Cantorian in 3-D would sweep out a 4-D world sheet; i.e. the time dimension is created by the Cantorian space $\mathfrak{E}^{\infty}$ (! - ?). Conjecturing further (wildly) one could say that perhaps space (and gravity) is created by the fractality of time. This is a typical form of conjecture to be found in the El Naschie papers - extremely thought provoking but ultimately heuristic. Regarding the Hurst exponent one recalls that for Feynmann trajectories in $1+1$ dimensions $(\mathbf{C14})\ d_{path} = 1/H = 1/d_c^{(0)} = d_c^{(2)}$. Thus we are concerned with relating **(C13)** and **(C14)** (among other matters). Note that path dimension is often thought of as a fractal dimension (M-B or box dimension), which is not necessarily the same as the Hausdorff dimension. However in [1] one shows that quantum mechanical free motion produces fractal paths of Hausdorff dimension 2 (cf. also [76]). ■

Remark 4.4. Following [25] let $S_c^{(0)}$ correspond to the set with dimension $d_c^{(0)} = \phi$. Then the complementary dimension is $\tilde{d}_c^{(0)} = 1-\phi = \phi^2$. The path dimension is gien as in **(C14)** by $(\mathbf{C15})\ d_{path} = d_c^{(2)} = 1/\phi = 1+\phi$ and $\tilde{d}_{path} = \tilde{d}_c^{(2)} = 1/(1-\phi) = 1/\phi^2 = (1+\phi)^2$. Following El Naschie for an equivalence between unions and intersections in a given space one requires (in the present situation) that

$$d_{crit} = d_c^{(2)} + \tilde{d}_c^{(2)} = \frac{1}{\phi} + \frac{1}{\phi^2} = \frac{\phi(1+\phi)}{\phi^3} = \frac{1}{\phi^3} = \frac{1}{\phi}\cdot\frac{1}{\phi^2} = d_c^{(2)}\cdot\tilde{d}_c^{(2)} = 4+\phi^3 \quad (4.6)$$

where $(\mathbf{C16})\ d_{crit} = 4+\phi^3 = d_c^{(4)} \sim 4.236$. Thus the critical dimension coincides with the Hausdorff dimension of $S_c^{(4)}$ which is embedded densely into a smooth space of topological dimension 4. On the other hand the backbone set of dimension $d_c^{(0)} = \phi$ is embedded densely into a set of topological dimension zero (a point). Thus one thinks in general of $d_c^{(n)}$ as the H dimension of a Cantor set of dimension ϕ embedded into a smooth space of integer topological dimension n. ■

Remark 4.5. In [25] it is also shown that realization of the spaces $\mathfrak{E}^{(n)}$ comprising $\mathfrak{E}^{\infty}$ can be expressed via the fractal sprays of Lapidus-van Frankenhuysen (cf. [80]). Thus we refer to [80] for graphics and details and simply sketch some ideas here (with apologies to M. Lapidus). A fractal string is a bounded open subset of **R** which is a disjoint union of an infinite number of open intervals $\mathfrak{L} = \ell_1, \ell_2, \cdots$. The geometric zeta function of $\mathfrak{L}$ is $(\mathbf{C17})\ \zeta_{\mathfrak{L}}(s) = \sum_1^{\infty} \ell_j^{-s}$. One assumes a suitable meromorphic extension of $\zeta_{\mathfrak{L}}$ and the complex dimensions of $\mathfrak{L}$ are defined as the poles of this meromorphic extension. The spectrum of $\mathfrak{L}$ is the sequence of frequencies $f = k\cdot\ell_j^{-1}$ $(k = 1, 2, \cdots)$ and the spectral zeta function of $\mathfrak{L}$ is defined as $(\mathbf{C18})\ \zeta_{\nu}(s) = \sum_f f^{-s}$ where in fact $\zeta_{\nu}(s) = \zeta_{\mathfrak{L}}(s)\zeta(s)$ (with $\zeta(s)$ the classical Riemann zeta function). Fractal sprays are higher dimensional generalizations of fractal strings. As an example consider the spray Ω obtained by scaling an open square B of size 1 by the lengths of the standard triadic Cantor string CS. Thus Ω consists of one open square of size 1/3, 2 open squares of size 1/9, 4 open squares of size 1/27, etc.

(see [80] for pictures and explanations). Then the spectral zeta function for the Dirichlet Laplacian on the square is $(\mathbf{C19})$ $\zeta_B(s) = \sum_{n_1,n_2=1}^{\infty}(n_1^2+n_2^2)^{s/2}$ and the spectral zeta function of the spray is $(\mathbf{C20})$ $\zeta_\nu(s) = \zeta_{CS}(s)\cdot\zeta_B(s)$. Now $\mathfrak{E}^\infty$ is composed of an infinite hierarchy of sets $\mathfrak{E}^{(j)}$ with dimension $(1+\phi)^{j-1} = 1/\phi^{j-1}$ $(j = 0, \pm 1, \pm 2, \cdots)$ and these sets correspond to a special case of boundaries $\partial\Omega$ for fractal sprays Ω whose scaling ratios are suitable binary powers of $2^{-\phi^{j-1}}$. Indeed for $n = 2$ the spectral zeta function of the fractal golden spray indicated above is $(\mathbf{C21})$ $\zeta_\nu(s) = (1/(1-2\cdot 2^{s\phi})\zeta_B(s)$. The poles of $\zeta_B(s)$ do not coincide with the zeros of the denominator $1-2\cdot 2^{-s\phi}$ so the (complex) dimensions of the spray correspond to those of the boundary $\partial\Omega$ of Ω. One finds that the real part $\Re s$ of the complex dimensions coincides with $dim\ \mathfrak{E}^{(2)} = 1+\phi = 1/\phi^2$ and one identifies then $\partial\Omega$ with $\mathfrak{E}^{(2)}$. The procedure generalizes to higher dimensions (with some stipulations) and for dimension n there results $\Re s = 1/\phi^{n-1} = dim\ \mathfrak{E}^{(n)}$. This produces a physical model of the Cantorian fractal space from the boundaries of fractal sprays (see [25] for further details and [80] for precision). Other (putative) geometric realizations of $\mathfrak{E}^\infty$ are indicated in [104] in terms of wild topologies, etc. ■

5. Hydrodynamics and the Fractal Schrödinger Equation

We sketch first some material from [3] (see also [14, 114, 115, 116] and Sections 2-4 for background). Thus let ψ be the wave function of a test particle of mass m_0 in a force field $U(r,t)$ determined via $(\mathbf{D1})$ $i\hbar\partial_t\psi = U\psi - (\hbar^2/2m)\nabla^2\psi$ where $\nabla^2 = \Delta$. One writes $(\mathbf{D2})$ $\psi(r,t) = R(r,t)exp(iS(r,t))$ with $(\mathbf{D3})$ $v = (\hbar/2m)\nabla S$ and $\rho = R\cdot R$ (one assumes $\rho \neq 0$ for physical meaning). Thus the field equations of QM in the hydrodynamic picture are

$$d_t(m_0\rho v) = \partial_t(m_0\rho v) + \nabla(m_0\rho v) = -\rho\nabla(U+Q);\ \partial_t\rho + \nabla\cdot(\rho v) = 0 \qquad (5.1)$$

where $(\mathbf{D4})$ $Q = -(\hbar^2/2m_0)(\Delta\sqrt{\rho}/\sqrt{\rho})$ is the quantum potential (or interior potential). Now because of the nondifferentiability of spacetime an infinity of geodesics will exist between any couple of points A and B. The ensemble will define the probability amplitude (this is a nice assumption but what is a geodesic here). At each intermediate point C one can consider the family of incoming (backward) and outgoing (forward) geodesics and define average velocities $b_+(C)$ and $b_-(C)$ on these families. These will be different in general and following Nottale this doubling of the velocity vector is at the origin of the complex nature of QM. Even though Nottale reformulates Nelson's stochastic QM the former's interpretation is profoundly different. While Nelson (cf. [111]) assumes an underlying Brownian motion of unknown origin which acts on particles in a still Minkowskian spacetime, and then introduces nondifferentiability as a byproduct of this hypothesis, Nottale assumes as a fundamental and universal principle that spacetime itself is no longer Minkowskian nor differentiable. While with Nelson's Browian motion hypothesis, nondifferentiability is but an approximation which expected to break down at the scale of the underlying collisions (?), where a new physics should be introduced, Nottale's hypothesis of nondifferentiability is essential and should hold down to the smallest possible length scales. (This sentence is interesting but needs elaboration). Following Nelson one defines now the mean forward and backward derivatives

$$\frac{d_\pm}{dt}y(t) = lim_{\Delta t \to 0_\pm}\left\langle \frac{y(t+\Delta t) - y(t)}{\Delta t}\right\rangle \tag{5.2}$$

This gives forward and backward mean velocities $(\mathbf{D5})$ $(d_+/dt)x(t) = b_+$ and $(d_-/dt)x(t) = b_-$ for a position vector x. Now in Nelson's stochastic mechanics one writes two systems of equations for the forward and backward processes and combines them in the end in a complex equation, Nottale works from the beginning with a complex derivative operator

$$\frac{\delta}{dt} = \frac{(d_+ + d_-) - i(d_+ - d_-)}{2dt} \tag{5.3}$$

leading to $(\mathbf{D6})$ $V = (\delta/dt)x(t) = v - iu = (1/2)(b_+ + b_-) - (i/2)(b_+ - b_-)$. One defines also $(\mathbf{D7})$ $(d_v/dt) = (1/2)(d_+ + d_-)/dt$ and $(d_u/dt) = (1/2)(d_+ - d_-)/dt$ so that $d_v x/dt = v$ and $d_u x/dt = u$. Here v generalizes the classical velocity while u is a new quantity arising from nondifferentiability. This leads to a stochastic process satisfying (respectively for the forward $(dt > 0)$ and backward $(dt < 0)$ processes) $(\mathbf{D8})$ $dx(t) = b_+[x(t)] + d\xi_+(t) = b_-[x(t)] + d\xi_-(t)$. The $d\xi(t)$ terms can be seen as fractal functions and they amount to a Wiener process when $\mathcal{D} = 2$ (presumably the fractal dimension). Then the $d\xi(t)$ are Gaussian with mean zero, mutually independent, and satisfy $(\mathbf{D9})$ $< d\xi_{\pm i}d\xi_{\pm j} >= \pm 2D\delta_{ij}dt$ where D is a diffusion coefficient. D can be found via $D = \hbar/2m_0$ given $\tau_0 = \hbar/(m_0c^2)$ (deBroglie time scale in the rest frame - cf [14] for more on this). Now **(D9)** allows one to give a general expression for the complex time derivative, namely

$$df = \frac{\partial f}{\partial t} + \nabla f \cdot dx + \frac{1}{2}\frac{\partial^2 f}{\partial x_i \partial x_j}dx_i dx_j \tag{5.4}$$

Next compute the forward and backward derivatives of f. Then $< dx_i dx_j > \to < d\xi_{\pm i}d\xi_{\pm j} >$ so the last term in (5.4) amounts to a Laplacian via **(D9)** and one obtains $(\mathbf{D10})$ $(d_\pm f/dt) = [\partial_t + b_\pm \cdot \nabla \pm D\Delta]f$. This is an important result. Thus assume the fractal dimension is not 2 in which case there is no longer a cancellation of the scale dependent terms in (5.4) and instead of $D\Delta f$ one would obtain an explicitly scale dependent behavior $D\delta t^{(2/D)-1}\Delta f$. In other words the value $D = 2$ implies that the scale symmetry becomes hidden in the operator formalism. Using **(D10)** one obtains the complex time derivative operator in the form $(\mathbf{D11})$ $(\delta/dt) = \partial_t + V \cdot \nabla - iD\Delta$ (cf. **(D6)** for V). Nottale's prescription is then to replace d/dt by δ/dt. In this spirit one can write now $(\mathbf{D12})$ $\psi = exp(i(\mathfrak{S}/2m_0D))$ so that $(\mathbf{D13})$ $V = -2iD\nabla(log(\psi))$ and then the generalized Newton equation $(\mathbf{D14})$ $-\nabla U = m_0(\delta/dt)V$ reduces to the SE.

Now assume the velocity field from the hydrodynamic model agrees with the real part v of the complex velocity V and equate the wave functions from the two models **(D12)** and **(D2)**; one obtains for $\mathfrak{S} = s + i\sigma$ $(\mathbf{D15})$ $s = 2m_0DS$, $D = (\hbar/2m_0)$, and $\sigma = -m_0Dlog(\rho)$. Using the definition $(\mathbf{D16})$ $V = (1/m_0)\nabla\mathfrak{S} = (1/m_0)\nabla s + (i/m_0)\nabla\sigma = v - iu$ (which results via **(D6)** by putting **(D12)** into **(D13)**) we get $(\mathbf{D17})$ $v = (1/m_0)\nabla s = 2D\nabla S$ and $u = -(1/m_0)\nabla\sigma = D\nabla log(\rho)$. Note that the imaginary part of the complex velocity given in **(D17)** coincides with Nottale. Dividing the time dependent SE **(D1)** by $2m_0$ and taking the gradient gives $(\mathbf{D18})$ $\nabla U/m_0 =$

$2D\nabla[i\partial_t log(\psi) + D(\Delta\psi/\psi)]$ where $\hbar/2m_0$ has been replaced by D. Then consider the identities

$$\Delta\nabla = \nabla\Delta;\ (\nabla f\cdot\nabla)(\nabla f) = (1/2)\nabla(\nabla f)^2;\ \frac{\Delta f}{f} = \Delta log(f) + (\nabla log(f))^2 \quad (5.5)$$

Now the second term in the right of **(D18)** becomes $(\mathbf{D19})$ $\nabla(\Delta\psi/\psi) = \Delta(\nabla log(\psi)) + 2(\nabla log(\psi)\cdot\nabla)(\nabla log(\psi))$ so **(D18)** can be written as $(\mathbf{D20})$ $\nabla U = 2iDm_0[\partial_t\nabla log(\psi) - iD\Delta(\nabla log(\psi) - 2iD(\nabla log(\psi)\cdot\nabla)(\nabla log(\psi))]$. One can show that **(D20)** is nothing but the generalized Newton equation **(D14)**. Now if we replace the complex velocity **(D13)**, taking into account **(D6)** and **(D17)** we get

$$-\nabla U = m_0\{\partial_t(v - iD\nabla log(\rho) + [i(v - iD\nabla log(\rho)\cdot\nabla](v - iD\nabla log(\rho)) - \quad (5.6)$$
$$-iD\Delta(v - iD\nabla log(\rho))\}$$

Equation (5.6) is a complex differential equation and reduces to (using (5.5))

$$m_0[\partial_t v + (v\cdot\nabla)v] = -\nabla\left(U - 2m_0D^2\frac{\Delta\sqrt{\rho}}{\sqrt{\rho}}\right);\ \nabla\left\{\frac{1}{\rho}\left[\partial_t\rho + \nabla\cdot(\rho v)\right]\right\} \quad (5.7)$$

The last equation in (5.7) reduces to the continuity equation up to a phase factor $\alpha(t)$ which can be set equal to zero (note again that $\rho \neq 0$ is posited). Thus (5.7) is nothing but the fundamental equations (5.1) of the hydrodynamic model. Further combining the imaginary part of the complex velocity in **(D17)** with the quantum potential **(D4)** and using (5.5) one gets $(\mathbf{D21})$ $Q = -m_0D\nabla\cdot u - (1/2)m_0u^2$. Since u arises from nondifferentiability according to our nondifferentiable space model of QM it follows that the quantum potential comes from the nondifferentiability of the quantum spacetime (very nice but where is $\mathfrak{E}^\infty$ from the title of [3] - also the x derivatives should be clarified).

Putting $U = 0$ in the first equation of (5.7), multiplying by ρ, and taking the second equation into account yields

$$\partial_t(m_0\rho\nu_k) + \frac{\partial}{\partial x_i}(m_0\rho\nu_i\nu_k) = -\rho\frac{\partial}{x_k}\left[2m_0D^2\frac{1}{\sqrt{\rho}}\frac{\partial}{\partial x_i}\frac{\partial}{\partial x_i}(\sqrt{\rho})\right] \quad (5.8)$$

(here $\nu_k \sim v_k$ seems indicated). Now set $(\mathbf{D22})$ $\Pi_{ik} = m_0\rho\nu_i\nu_k - \sigma_{ik}$ along with $\sigma_{ik} = m_0\rho D^2(\partial/\partial x_i)(\partial/\partial x_k)(log(\rho))$. Then (5.8) takes the simple form $(\mathbf{D23})$ $\partial_t(m_0\rho\nu_k) = -\partial\Pi_{ik}/\partial x_i$. The analogy with classical fluid mechanics works well if one introduces the kinematic $(\mathbf{D24})$ $\mu = D/2$ and dynamic $\eta = (1/2)m_0D\rho$ viscosities. Then Π_{ik} defines the momentum flux density tensor and σ_{ik} the internal stress tensor $(\mathbf{D25})$ $\sigma_{ik} = \eta[(\partial u_i/\partial x_k) + (\partial u_k/\partial x_i)]$. From **(D22)** one can see that the internal stress tensor is build up using the quantum potential while the equations (5.1) or (5.7) are nothing but systems of Navier-Stokes type for the motion where the quantum potential plays the role of an internal stress tensor. In other words the nondifferentiability of the quantum spacetime manifests itself like an internal stress tensor. For clarity in understanding **(D23)** we put this in one dimensional form so (5.8) becomes

$$\partial_t(m_0\rho v) + \partial_x(m_0\rho v^2) = -\rho\partial\left(2m_0D^2\frac{1}{\sqrt{\rho}}\partial^2\sqrt{\rho}\right) = \rho\partial Q \quad (5.9)$$

and $\Pi = m_0\rho v^2 - \sigma$ with $\sigma = m_0\rho D^2\partial^2 log(\rho)$. This agrees in the standard formulas (cf. [14]). Now note $\partial\sqrt{\rho} = (1/2)\rho^{-1/2}\rho'$ and $\partial^2\sqrt{\rho} = (1/2)[-(1/2)\rho^{-3/2}(\rho')^2 + \rho^{-1/2}\rho'']$ with $\partial^2 log(\rho) = \partial(\rho'/\rho) = (\rho''/\rho) - (\rho'/\rho)^2$ while

$$-\rho\partial\left[2m_0D^2\frac{1}{\sqrt{\rho}}\left(\partial^2\sqrt{\rho}\right)\right] = -2m_0D^2\rho\partial\left[\frac{1}{2\sqrt{\rho}}\left(-\frac{1}{2}\rho^{-3/2}(\rho')^2 + \rho^{-1/2}\rho''\right)\right] = \tag{5.10}$$

$$= -2m_0D^2\rho\partial\left[\frac{\rho''}{2\rho} - \frac{1}{4}\left(\frac{(\rho'}{\rho}\right)^2\right] = -m_0D^2\rho\partial\left[\frac{\rho''}{\rho} - \frac{1}{2}\left(\frac{\rho'}{\rho}\right)^2\right]$$

One wants to show then that **(D23)** holds or equivalently $-\partial\sigma = (5.10)$. Here

$$-\partial\sigma = -\partial[m_0\rho D^2\partial^2 log(\rho)] = -m_0D^2\left[\rho'\left(\frac{\rho''}{\rho} - \left(\frac{\rho'}{\rho}\right)^2\right) + \rho\partial\left(\frac{\rho''}{\rho} - \frac{(\rho')^2}{\rho}\right)\right] \tag{5.11}$$

so we want (5.11) = (5.10) and this is easily verified.

6. Recapitulation

We write down now some of the main formulas here (with some unification of notation) in order to help provide perspective. The goal is not entirely clear but many questions will arise as we go along and at the end. Hopefully we will be able to answer some of the questions.

1. We write from Section 2 (**E1**) $\psi = Rexp(iS/\hbar)$ with

$$S_t + \frac{(S')^2}{2m} + V - \frac{\hbar^2}{2m}\frac{R''}{R} = 0;\ \partial_t(R^2) + \frac{1}{m}(R^2S')' = 0 \tag{6.1}$$

For $P = R^2$ and $Q = -(\hbar^2/2m)(R''/R)$ this yields

$$S_t + \frac{(S')^2}{2m} + Q + V = 0;\ P_t + \frac{1}{m}(PS')' = 0 \tag{6.2}$$

Writing $\rho = mP$ and $p = m\dot{x}$ leads to

$$\partial_t(\rho v) + \partial(\rho v^2) + \frac{\rho}{m}\partial V - \frac{\hbar^2}{2m^2}\rho\partial\left(\frac{\partial^2\sqrt{\rho}}{\sqrt{\rho}}\right) = 0 \tag{6.3}$$

Along the way one arrived at (2.8) and "completed" this with a pressure term $\nabla F = \rho^{-1}\nabla\mathfrak{p}$ or $F' = (1/R^2)\mathfrak{p}'$ to arrive at (**E2**) $mv_t + mvv' = -\partial(V+Q) - F'$ corresponding to a SE (**E3**) $i\hbar\psi_t = -(\hbar^2/2m)\psi'' + V\psi + F\psi$. One wants then $F = F(\psi)$.

2. Consider a quantum state corresponding to a "subquantum" statistical ensemble governed by classical kinetics in a phase space. One arrives at $\psi = \rho^{1/2}exp(i\mathfrak{S}/\hbar)$ with (**E4**) $i\hbar\psi_t = -(\hbar^2/2m)\psi_{xx} + \mathcal{V}\psi$ where $\mathfrak{S} = NS$, $N = \int|\psi|^2 d^nx$, $\hbar =$

$N\eta$, $m = N\mu$, $\mathcal{V} = NV$, and $log(\psi) = (1/2)log(\rho) + (i/\eta)S$. The fields ρ, S or ξ, S determine a quantum fluid with (cf. (2.15))

$$\frac{\partial \xi}{\partial t} + \frac{1}{\mu}\frac{\partial^2 S}{\partial x^2} + \frac{1}{\mu}\frac{\partial \xi}{\partial x}\frac{\partial S}{\partial x} = 0; \tag{6.4}$$

$$\frac{\partial S}{\partial t} - \frac{\eta^2}{4\mu}\frac{\partial^2 \xi}{\partial x^2} - \frac{\eta^2}{8\mu}\left(\frac{\partial \xi}{\partial x}\right)^2 + \frac{1}{2\mu}\left(\frac{\partial S}{\partial x}\right)^2 + V = 0$$

which for $\psi = \rho^{1/2}exp(i\mathfrak{S}/\hbar)$ leads to

$$i\hbar\frac{\partial \Psi}{\partial t} = -\frac{\hbar^2}{2m}\frac{\partial^2 \Psi}{\partial x^2} + \mathcal{V}\Psi \tag{6.5}$$

3. The Fisher information connection ła Remarks 2.4-2.5 involves a classical ensemble with particle mass m moving under a potential V

$$S_t + \frac{1}{2m}(S')^2 + V = 0;\ P_t + \frac{1}{m}\partial(PS')' = 0 \tag{6.6}$$

where S is a momentum potential; note that no quantum potential is present but this will be added on in the form of a term $(1/2m)\int dt(\Delta N)^2$ in the Lagrangian which measures the strength of fluctuations. This can then be specified in terms of the probability density P as indicated in Remark 2.4 leading to a SE (2.24). A "neater" approach is given in Remark 2.5 leading in 1-D to

$$S_t + \frac{1}{2m}(S')^2 + V + \frac{\lambda}{m}\left(\frac{(P')^2}{P^2} - \frac{2P''}{P}\right) = 0 \tag{6.7}$$

Note that $Q = -(\hbar^2/2m)(R''/R)$ becomes for $R = P^{1/2}$ (**E5**) $Q = -(2\hbar^2/2m)[(2P''/P) - (P'/P)^2]$ (cf. (2.4)). Thus the addition of the Fisher information serves to quantize the classical system.

4. One defines an information entropy (IE) in Remark 2.6 via (**E6**) $\mathfrak{S} = -\int \rho log(\rho)d^3x$ $(\rho = |\psi|^2)$ leading to

$$\frac{\partial \mathfrak{S}}{\partial t} = \int (1 + log(\rho))\partial(v\rho) \sim \int \frac{(\rho')^2}{\rho} \tag{6.8}$$

modulo constants involving $D \sim \hbar/2m$. $\mathfrak{S}$ is typically not conserved and $\partial_t\rho = -\nabla\cdot(v\rho)$ $(u = D\nabla log(\rho)$ with $v = -u$ corresponds to standard Brownian motion with $d\mathfrak{S}/dt \geq 0$. Then high IE production corresponds to rapid flattening of the probability density. Note here also that $\mathfrak{F} \sim -(2/D^2)\int \rho Q dx = \int dx[(\rho')^2/\rho]$ is a functional form of Fisher information. Entropy balance is discussed in Remark 2.8 and the manner in which Q appears in the hydrodynamical formalism is exhibited in (2.38)-(2.39).

5. The Nagasawa theory (based in part on Nelson's work) is very revealing and fascinating (see [90, 91]). The essense of Theorem 3.1 is that $\psi = exp(R + iS)$ satisfies the SE (**E7**) $i\psi_t + (1/2)\psi'' + ia\psi' - V\psi = 0$ if and only if

$$V = -S_t + \frac{1}{2}R'' + \frac{1}{2}(R')^2 - \frac{1}{2}(S')^2 - aS;\ 0 = R_t + \frac{1}{2}S'' + S'R' + aR' \quad (6.9)$$

Changing variables in (**E8**) ($X = (\hbar/\sqrt{m})x$ and $T = \hbar t$) one arrives at (**E9**) $i\hbar\psi_T = -(\hbar^2/2m)\psi_{XX} - iA\psi_X + V\psi$ where $A = a\hbar/\sqrt{m}$ and

$$i\hbar R_T + (\hbar^2/m^2)R_X S_X + (\hbar^2/2m^2)S_{XX} + AR_X = 0; \quad (6.10)$$

$$V = -i\hbar S_T + (\hbar^2/2m)R_{XX} + (\hbar^2/2m^2)R_X^2 - (\hbar^2/2m^2)S_X^2 - AS_X$$

The diffusion equations then take the form

$$\hbar\phi_T + \frac{\hbar^2}{2m}\phi_{XX} + A\phi_X + \tilde{c}\phi = 0;\ -\hbar\hat{\phi}_T + \frac{\hbar^2}{2m}\hat{\phi}_{XX} - A\hat{\phi}_X + \tilde{c}\hat{\phi} = 0; \quad (6.11)$$

$$\tilde{c} = -\tilde{V}(X,T) - 2\hbar S_T - \frac{\hbar^2}{m}S_X^2 - 2AS_X$$

It is now possible to introduce a role for the quantum potential in this theory. Thus from $\psi = exp(R + iS)$ (with $\hbar = m = 1$ say) we have $\psi = \rho^{1/2}exp(iS)$ with $\rho^{1/2} = exp(R)$ or $R = (1/2)log(\rho)$. Hence $(1/2)(\rho'/\rho) = R'$ and $R'' = (1/2)[(\rho''/\rho) - (\rho'/\rho)^2]$ while the quantum potential is $Q = (1/2)(\partial^2\rho^{1/2}/\rho^{1/2}) = -(1/8)[(2\rho''/\rho) - (\rho'/\rho)^2]$ (cf. (2.4)). Equation (6.9) becomes then

$$V = -S_t + \frac{1}{8}\left(\frac{2\rho''}{\rho} - \frac{(\rho')^2}{\rho^2}\right) - \frac{1}{2}(S')^2 - aS \equiv S_t + \frac{1}{2}(S')^2 + V + Q + aS = 0; \quad (6.12)$$

$$\rho_t + \rho S'' + S'\rho' + a\rho' = 0 \equiv \rho_t + (\rho S')' + a\rho' = 0$$

Thus $-2S_t - (S')^2 = 2V + 2Q + 2AS$ and one has

Proposition 6..1. The creation-annihilation term c in the diffusion equations (cf. Theorem 3.1) becomes

$$c = -V - 2S_t - (S')^2 - 2aS' = V + 2Q + 2a(S - S') \quad (6.13)$$

where Q is the quantum potential.

6. Regarding scale relativity one writes (cf. (3.12)

$$\frac{d_\pm}{dt}y(t) = lim_{\Delta t \to 0\pm}\left\langle\frac{\pm y(t \pm \Delta t) \mp y(t)}{\Delta t}\right\rangle \quad (6.14)$$

and we collect equations in ($\rho = |\psi|^2$)

$$dx = b_+dt + d\xi_+ = b_-dt + d\xi_-; < d\xi_+^2 >= 2\mathcal{D}dt = - < d\xi_-^2 > \quad (6.15)$$

$$\frac{d_+f}{dt} = (\partial_t + b_+\partial + \mathcal{D}\partial^2)f;\ \frac{d_-f}{dt} = (\partial_t + b_-\partial - \mathcal{D}\partial^2)f \quad (6.16)$$

$$V = \frac{1}{2}(b_+ + b_-);\ U = \frac{1}{2}(b_+ - b_-);\ \rho_t + \partial(\rho V) = 0;\ U = \mathcal{D}\partial(log(\rho));\quad (6.17)$$

$$\mathcal{V} = V - iU;\ d_{\mathcal{V}} = \frac{1}{2}(d_+ + d_-);\ d_{\mathcal{U}} = \frac{1}{2}(d_+ - d_-)$$

$$\frac{d_{\mathcal{V}}}{dt} = \partial_t + V\partial;\ \frac{d_{\mathcal{U}}}{dt} = \mathcal{D}\partial^2 + U\partial;\ \frac{d'}{dt} = (\partial_t - i\mathcal{D}\partial^2) + \mathcal{V}\partial \quad (6.18)$$

$$V = 2\mathcal{D}\partial S;\ \mathcal{S} = log(\rho^{1/2}) + iS;\ \psi = \sqrt{\rho}e^{iS} = e^{i\mathcal{S}};\ \mathcal{V} = -2i\mathcal{D}\partial log(\psi) \quad (6.19)$$

For Lagrangian $\mathcal{L} = (1/2)m\mathcal{V}^2 - m\mathfrak{U}$ one gets a SE

$$i\hbar\psi_t = -\frac{\hbar^2}{2m}\partial^2\psi + \mathfrak{U}\psi \quad (6.20)$$

coming from Newton's law (**E10**) $-\partial\mathfrak{U} = -2i\mathcal{D}m(d'/dt)\partial log(\psi) = m(d'/dt)\mathcal{V}$.

7. The development in Section 3 based on [23] involves thinking of nonlinear QM as a fractal Brownian motion with complex diffusion coefficient. We note **(E10)** corresponds to **(B58)** and **(B55)** arises in (6.18). These give rise to

$$-\nabla U = -2im[D\partial_t\nabla log(\psi)] - 2D\nabla\left(D\frac{\nabla^2\psi}{\psi}\right) \quad (6.21)$$

Thus putting in a complex diffusion coefficient leads to the NLSE

$$i\hbar\partial_t\psi = -\frac{\hbar^2}{2m}\frac{\alpha}{\hbar}\nabla^2\psi + U\psi - i\frac{\hbar^2}{2m}\frac{\beta}{\hbar}(\nabla log(\psi))^2\psi \quad (6.22)$$

with $\hbar = \alpha + i\beta = 2mD$ complex.

8. In [3] one writes again $\psi = Rexp(iS/\hbar)$ with field equations in the hydrodynamical picture

$$d_t(m_0\rho v) = \partial_t(m_0\rho v) + \nabla(m_0\rho v) = -\rho\nabla(u + Q);\ \partial_t\rho + \nabla\cdot(\rho v) = 0 \quad (6.23)$$

where $Q = -(\hbar^2/2m_0)(\Delta\sqrt{\rho}/\sqrt{\rho})$. One works with the Nottale approach as above with $d_v \sim d_{\mathcal{V}}$ and $d_u \sim d_{\mathcal{U}}$ (cf. (6.18)). One assumes that the velocity field from the hydrodynamical model agrees with the real part v of the complex velocity $V = v - iu$ so (cf. (6.17)) $v = (1/m_0)\nabla s \sim 2D\partial s$ and $u = -(1/m_0)\nabla\sigma \sim D\partial log(\rho)$ where $D = \hbar/2m_0$. In this context the quantum potential $Q = -(\hbar^2/2m_0)\Delta\sqrt{\rho}/\sqrt{\rho}$ becomes (**E11**) $Q = -m_0D\nabla u - (1/2)m_0u^2 \sim -(\hbar/2)\partial u - (1/2)m_0u^2$. Consequently Q arises from the fractal derivative and the nondifferentiability of spacetime. Further one can relate u (and hence Q) to an internal stress tensor **(D25)** whereas the v equations correspond to systems of Navier-Stokes type. Note here that (5.9) involves a term relating the stress tensor Π and Q directly.

7. Conclusions

One feature either exhibited or suggested in the examples displayed involves the role of a quantum potential in either quantization or "classicalization" of certain systems of equations of hydrodynamic type. Now with numbers referring to Section 6 we have:

1. One arrived at an equation of hydrodynamic type directly from the SE upon addition of a pressure term which served to augment the original potential V (however this could have simply been included in V). On the other hand Q does not appear in the SE but is generated by the decomposition $\psi = Rexp(iS/\hbar)$

2. In a general statistical mechanical approach, with the dynamics determined by classical kinetics in a phase space, the quantum potential has an interpretation in terms of an internal stress tensor for a quantum fluid. The equations are again described in terms of a probability density ρ and a phase factor S.

3. In #3-#4 one takes a classical statistical ensemble with S a momentum potential and expresses momentum fluctuations in terms of Fisher information; this leads to a SE with quantization term Q expressed as Fisher information. In Remarks 2.6-2.8 we show how Fisher information, entropy, and the quantum potential are mutually entangled (cf. also [39]). In (2.38)-(2.39) (based on [55]) we see how the Euler equation $(\partial_t + v \cdot \nabla)v = (F/m) - (\Delta P/\rho)$ (where P is a pressure term) is related to the quantized form $(\partial_t + v \cdot \nabla)v = (F/m) - \nabla Q$ arising from a SE.

4. The Nagasawa-Nelson approach in #5 views matters rather differently in showing the equivalence of the SE to a pair of diffusion equations. The full theory is very elegant and extends to singular situations, etc. (cf. [90, 92]). It would be of interest here to further examine the quantum potential in this context.

5. In #6-#8 one arrives at a pair of equations by virtue of the "fractal" structure of space (where fractal here simply means that nondifferentiable paths are considered which generate a complex velocity). In [3] (as exhibited in #8) one relates the quantum potential to the velocity u, showing its origin in the "fractal" derivative idea.

We emphasize that in fact the quantum potential comes up in a serious manner in the Bohm theory, with refinements as in [11, 18, 19, 20, 21, 22, 43, 44, 45, 47, 48, 50, 51, 65, 66]. In fact, given that trajectories are at the base of this theory one can forsee a fractal Bohm theory in the future (cf. [67, 112]). On the other hand one can make convincing arguments for fields as the fundamental objects (except perhaps in the Bohmian type theories) with particules "emerging" (cf. [64, 142]) as in quantum field theory (or perhaps via ripples or fractal structure in spacetime itself).

It is not entirely clear how to handle derivatives in statistical or fractal theories. There are of course many powerful techniques available for Brownian motion and stochastic differential equations and there is a developing literature about differential calculus on fractals. Random walks and general discretization methods are also useful. Somehow one would like to imagine that the formal power of calculus (and duality via distribution like theories) might be strong enough to override the microscopic details about the domains of differential operators. Perhaps the coordinate derivative operators in situations such as #6-#8 could be defined so that their domains are various fractal sets densely embedded in $\mathbf{R}^n$ (in this connection see e.g. [30, 74, 75, 87, 110, 123, 125, 137]). In the end the most attractive formulation would seem to be some (more or less rigorous) version of a Feynmann path integral where precise definitions of the path space are not critical.

References

[1] L. Abbott and M. Wise, *Amer. Jour. Physics*, **49** (1981), 37-39

[2] F. Ben Adda and J. Cresson, Quantum derivatives and the Schrödinger equation, *Chaos, Solitons, and Fractals*, **19** (2004), 1323-1334

[3] M. Agop, P. Ioannou, C. Buzea, and P. Nica, *Chaos, Solitons, and Fractals*, **16** (2003), 321-338

[4] J. Almeida, *Hamiltonian systems: Chaos and quantization*, Cambridge Univ. Press, 1988

[5] M. Barnsley, *Fractals everywhere*, Academic Press, 1988

[6] F. Berezin and M. Shubin, *The Schrödinger equation*, Kluwer, 1991

[7] J. Berger, quant-ph 0309143

[8] G. Bertoldi, A. Faraggi, and M. Matone, hep-th 9909201

[9] I. Bialynicki-Birula and J. Mycielski, *Annals Phys.*, **100** (1976), 62

[10] A. Boyarsky and P. Gora, *Chaos, Solitons, and Fractals*, **9** (2001), 1611-1618; 7 (1996), 611-630, 939-954

[11] R. Carroll, *Quantum theory, deformation, and integrability*, North-Holland, 2000

[12] R. Carroll, *Calculus revisited*, Kluwer, 2002

[13] R. Carroll, *Lecture notes on aspects of the Schrödinger equation* (2003 - 45 pages)

[14] R. Carroll, *Lecture notes on quantum theory, diffusion, hydrodynamics, and fractals*, (2003 - 46 pages)

[15] R. Carroll, *Some remarks on a fractal spacetime*, in preparation

[16] R. Carroll, *Lecture notes on uncertainty, trajectories, and quantum geometry* (2003 - 62 pages)

[17] R. Carroll, *Lecture notes on integrable systems, quantum mechanics, and q-theories* (2003 - 59 pages)

[18] R. Carroll, *Proc. Conf. Symmetry*, Kiev, 2003, to appear

[19] R. Carroll, *Canadian Jour. Phys.*, **77** (1999), 319-325

[20] R. Carroll, *Direct and inverse problems of mathematical physics* , Kluwer, 2000, pp. 39-52

[21] R. Carroll, *Generalized analytic functions*, Kluwer, 1998, pp. 299-311

[22] R. Carroll, quant-ph 0309223 and 0309159

[23] C. Castro, J. Mahecha, and B. Rodriguez, quant-ph 0202026

[24] C. Castro, *Chaos, Solitons, and Fractals*, **11** (2000), 1663-1670; 12 (2001), 101-104, 1585-1606

[25] C. Castro, hep-th 9512044, 0203086

[26] M. Célérier and L. Nottale, hep-th 0112213 and 0210027 Scientific, 1996

[27] K. Chung and Z. Zhao, *From Brownian motion to Schrödinger's equation* , Springer, 2001

[28] K. Chung and J. Zambrini, *Introduction to random time and quantum randomness* , World Scientific, 2003

[29] K. Chung and R. Williams, *Introduction to stochastic integration* , Birkhäuser, 1990

[30] A. Compte, *Phys. Rev. E*, **53** (1996), 4191-4193

[31] A. Connes, *Noncommutative geometry*, Academic Press, 1994

[32] R. Cremona and J. Lacroix, *Spectral theory of random Schrödinger*

[33] J. Cresson, math.GM 0211071

[34] J. Cresson, *Scale calculus and the Schrödinger equation; Scale geometry, I* , preprints 2003

[35] J. Cresson, *Nondifferentiable variational principles* , preprint 2003

[36] H. Cycon, R. Froese, W. Kirsch, and B. Simon, *Schrödinger operators with applications to quantum mechanics and global geometry* , Springer, 1987

[37] M. Czachor and H. Doebner, quant-ph 0106051 and 0110008

[38] R. Czopnik and P. Garbaczewski, quant-ph 0203018; cond-mat 0202463

[39] M. Davidson, quant-ph 0110050 and 0112157

[40] M. Davidson, quant-ph 0106124

[41] D. Delphenich, gr-qc 0211065

[42] H. Doebner, G. Goldin, and P. Natterman, quant-ph 9502014 and 9709036

[43] H. Doebner and G. Goldin, *Phys. Rev. A*, **54** (1996), 3764-

[44] D. Dürr, S. Goldstein, and N. Zanghi, quant-ph 9511016 and 0308039

[45] D. Dürr, S. Goldstein, and N. Zanghi, quant-ph 0308038

[46] K. Falconer, *Fractal geometry*, Wiley, 1990; *The geometry of fractal sets*, 1988; *echniques in fractal geometry*, Wiley, 1997

[47] A. Faraggi and M. Matone, *Phys. Rev. Lett.*, **78** (1997), 163-166

[48] A. Faraggi and M. Matone, *Inter. Jour. Mod. Phys. A*, 15 (2000), 1869-2017

[49] J. Fenyes, *Zeit. d. Phys.*, **132** (1952), 81-106

[50] E. Floyd, *Inter. Jour. Mod. Phys. A*, **14** (1999), 1111-1124; 15 (2000), 1363-1378; *Found. Phys. Lett.*, **13** (2000), 235-251; quant-ph 0009070, 0302128 and 0307090

[51] E. Floyd, *Phys. Rev. D*, **29** (1984) 1842-1844; 26 (1982), 1339-1347; 34 (1986), 3246-3249; 25 (1982), 1547-1551; *Jour. Math. Phys.*, **20** (1979), 83-85; 17 (1976), 880-884; *Phys. Lett. A*, **214** (1996), 259-265; *Inter. Jour. Theor. Phys.*, **27** (1998), 273-281

[52] B. Frieden, *Physics from Fisher information*, Cambridge Univ. Press, 1998

[53] B. Frieden, A. Plastino, A.R. Plastino, and B. Soffer, cond-mat 0206107

[54] T. Fülöp and S. Katz, quant-ph 9806067

[55] P. Garbaczewski, cond-mat 0211362 and 0301044

[56] G. Goldin and V. Shtelen, quant-ph 0006067

[57] I. Gottlieb, G. Ciobanu, and C. Buzea, *Chaos, Solitons, and Fractals*, **17**

[58] I. Gottlieb, M. Agop, and M. Jarcau, *Chaos, Solitons, and Fractals*, **19** (2004), 705-730

[59] J. Gouyet, *Physique et structures fractales*, Masson, 1992

[60] G. Grössing, quant-ph 0311109

[61] M. Hall, quant-ph 9806013, 9903045, 9912055, 0103072, 0107149, 0302007

[62] M. Hall and M. Reginatto, quant-ph 0102069 and 0201084

[63] M. Hall, K. Kumar, and M. Reginatto, hep-th 0206235

[64] H. Halvorson and R. Clifton, quant-ph 0103041

[65] P. Holland, *The quantum theory of motion*, Cambridge Univ. Press, 1997

[66] P. Holland, *Foundations of Physics*, **38** (1998), 881-911; *Nuovo Cimento B*, 116 (2001), 1043 and 1143

[67] R. Hyman, S. Caldwell, and E. Dalton, quant-ph 0401008

[68] D. Johnston, W. Janke, and R. Kenna, cond-mat 0308316

[69] M. Kaku, *Introduction to superstrings and M theory*, Springer, 1999; *Strings, conformal fields, and M theory*, Springer, 2000

[70] G. Kälbermann, quant-ph 0307018

[71] G. Kaniadakis and A. Scarfone, *Rep. Math. Phys.*, 51 (2003), 225 (cond-mat 0303334); *Jour. Phys. A*, **35** (2002), 1943 (quant-ph 0202032)

[72] G. Kaniadakis, *Physica A*, **307** (2002), 172 (quant-ph 0112049)

[73] G. Kaniadakis, *Phys. Lett. A*, **310** (2003), 377 (quant-ph 0303159); *Found. Phys. Lett.*, 16 (2003),99 (quant-ph 0209033)

[74] J. Kigami, *Analysis on fractals*, Cambridge Univ. Press, 2001

[75] K. Kolwankar and A. Gangal, cond-mat 9801138; physics 9801010

[76] H. Kröger, *Phys. Rev. A*, **55** (1997), 951-966

[77] G. Landi, *An introduction to noncommutative spaces and their geometry*, Springer, 1997

[78] N. Landsman, *Mathematical topics between classical and quantum mechanics*, Springer, 1998

[79] N. Lemos, *Phys. Lett. A*, **78** (1980), 237 and 239

[80] M. Lapidus and M. van Frankenhuysen, *Fractal geometry and number theory*, Birkhäuser, 2000

[81] R. Libof, *Kinetic theory*, Springer, 2003

[82] J. Madore, *An introduction to noncommutative geometry and its physical applications*, Cambridge Univ. Press, 1995

[83] S. Majid, *Foundations of quantum group theory*, Cambridge Univ. Press, 1995

[84] Yu. Manin, *Frobenius manifolds, quantum cohomology, and moduli spaces*, Amer. Math. Soc., 1999

[85] R. Mauldin and C. Williams, *Trans. Amer. Math. Soc.*, **295** (1986), 325-346

[86] A. Messiah, *Quantum mechanics*, Dover, 1999

[87] R. Metzler, E. Barkai, and J. Klafter, *Phys. Rev. Lett.*, **82** (1999), 3563-3567

[88] M. Nagasawa, *Prob. Theory Rel. Fields*, **82** (1089), 109-136; *Chaos, Solitons, and Fractals*, **7** (1996), 631-643

[89] M. Nagasawa, Stochastic processes, *Physics, and Geometry II*, World Scientific, 1995, pp. 545-556

[90] M. Nagasawa, *Schrödinger equations and diffusion theory* , Birkäuser, 1993

[91] M. Nagasawa, *Diffusion processes and related problems in analysis, I* , Birkhäuser, 1990, pp. 155-200

[92] M. Nagasawa, *Stochastic processes in quantum physics* , Birkhäuser, 2000

[93] M. El Naschie, *Chaos, Solitons, and Fractals* , **3** (1993), 89-98; 7 (1996), 499-518; 11 (1997), 1873-1886

[94] M. El Naschie, *Nuovo Cimento B*, **107** (1992), 583-594; *Chaos, Solitons, and Fractals*, 2 (19920, 91-94

[95] M. EL Naschie, *Chaos, Solitons, and Fractals* , **3** (1993), 675-685

[96] M. El Naschie, *Nuovo Cimento B*, **109** (1994), 149-157; *Chaos, Solitons, and Fractals*, **4** (1994) 177-179, 293-296, 2121-2132, 2269-2272

[97] M. El Naschie, *Chaos, Solitons, and Fractals* , **5** (1995), 661-684, 1503-1508; 6 (1995), 1031-1032

[98] M. El Naschie, *Chaos, Solitons, and Fractals* , **7** (1996), 955-959, 1501-1506

[99] M. El Naschie, *Chaos, Solitons, and Fractals* , **8** (1997), 753-759, 1865-1872

[100] M. El Naschie, *Chaos, Solitons, and Fractals* , **9** (1998), 913-919, 2023-2030

[101] M. El Naschie, *Chaos, Solitons, and Fractals* , **10** (1999), 567-580

[102] M. El Naschie, *Chaos, Solitons, and Fractals* , **11** (2000), 453-464, 2391-2395

[103] M. El Naschie, *Chaos, Solitons, and Fractals* , **12** (2001), 851-858

[104] M. El Naschie, *Chaos, Solitons, and Fractals* , **13** (2002), 1935-1945

[105] M. El Naschie, *Chaos, Solitons, and Fractals* , **18** (2003), 401-420

[106] M. El Naschie, *Chaos, Solitons, and Fractals* , **19** (2004), 209-236 and 689-697; 20 (2004), 437-450

[107] P. Natterman and W. Scherer, quant-ph 9506033

[108] P. Natterman and R. Zhdanov, solv-int 9510001

[109] P. Natterman, quant-ph 9703017 and 9709044

[110] J. Needleman, R. Strichartz, A. Teplyaev, and P. Yung, math.GM 0312027

[111] E. Nelson, *Quantum fluctuations*, Princeton Univ. Press, 1985; *Dynamical theory of Brownian motion*, Princeton Univ. Press, 1967

[112] H. Nikolic, quant-ph 0208185, 0302152, and 0307179; *Phys. Lett. B*, **527** (2002), 119-124

[113] L. Nottale, M. Célérier, and T. Lehner, hep-th 0307093

[114] L. Nottale, *Fractal space-time and microphysics: Toward a theory of scale relativity*, World Scientific, 1993

[115] L. Nottale, *Chaos, solitons, and Fractals*, **10** (1999), 459-468

[116] L. Nottale, *Chaos, Solitons, and Fractals*, **16** (2003), 539-564

[117] B. Oksendal, *Stochastic differential equations*, Springer, 2003

[118] L. Olavo, qunt-ph 9503020, 9503021, 9503022, 9503024, 9503025, 9509012, 9509013, 9511028, 9511039, 9601002, 9607002, 9607003, 9609003,9609023, 9703006, 9704004

[119] G. Ord, *Chaos, Solitons, and Fractals*, **8** (1997), 727-741; 9 (1998), 1011-1029; *Jour. Phys. A*, **16** (1983), 1869-1884

[120] G. Ord, *Chaos, Solitons, and Fractals*, **11** (2000), 383-391; 17 (2003), 609-620

[121] G. Ord and J. Gualtieri, *Chaos, Solitons, and Fractals*, **14** (2002), 929-935

[122] G. Ord and R. Mann, quant-ph 0206095; 0208004

[123] G. Ord and A. Deakin, *Phys. Rev. A*, **54** (1996), 3772-3778

[124] M. Pardy, quant-ph 0111105

[125] A. Parvate and A. Gangal, math-ph 0310047

[126] M. Pavon, *Jour. Math. Phys.*, **36** (1995), 6774-6800; quant-ph 0306052

[127] Y. Pesin, *Dimension theory in dynamical systems*, Univ. Chicago Press, 1997

[128] J. Polchinski, *String theory*, Vols. 1 and 2, Cambridge Univ. Press, 1998

[129] W. Puszkarz, quant-ph 9912006

[130] W. Puszkarz, quant-ph 9802001, 9903010, and 9905046

[131] W. Puszkarz, quant-ph 9710007, 0710008, 9710009, 9710010, and 9710011

[132] M. Reginatto, quant-ph 9909065

[133] F. Reif, *Fundamentals of statistical and thermal physics*, McGraw-Hill, 1965

[134] D. Revuz and M. Yor, *Continuous martingales and Brownian motion*, Springer, 1999

[135] C. Sabot, math-ph 0201041

[136] G. Sewell, *Quantum mechanics and its emergent macrophysics*, Princeton Univ. Press, 2002

[137] I. Sokolov, A. Chhechkin, and J. Klafter, cond-mat 0401146

[138] D. Stroock, *Markov processes from K. Ito's perspective*, Princeton Univ. Press, 2003

[139] C. Tricot, *Courbes et dimension fractale*, Springer, 1999

[140] P. Van and T. Fülöp, quant-ph 0304062

[141] P. Van, cond-mat 0112214 and 0210402

[142] D. Wallace, quant-ph 0112148 and 0112149

[143] B. West, M. Bologna, and P. Grigolini, *Physics of fractal operators*, Springer, 2003

[144] J. Zambrini, *Phys. Rev. A*, **38** (1987), 3631-3649; 33 (1986), 1532-1548; *Jour. Math. Phys.*, **27** (1986), 2307-2330

In: Focus on Evolution Equations
Editor: Gaston M. N'Guerekata

ISBN: 978-1-60021-342-7

Chapter 3

EXISTENCE OF p-ALMOST AUTOMORPHIC MILD SOLUTION TO SOME ABSTRACT DIFFERENTIAL EQUATIONS*

Toka Diagana†
Department of Mathematics, Howard University,
N.W - Washington, D.C., U.S.A.

Abstract

This paper is concerned with the properties of the class of p-almost automorphic functions recently introduced by the author as well as their applications to abstract differential equations of the form

$$u'(t) = Au(t) + f(t) \quad (*),$$

where A is the infinitesimal generator of a c_0-semigroup $(T(t))_{t\geq 0}$ in a Banach space $\mathbb{X}$ and $f : \mathbb{R} \mapsto \mathbb{X}$ is p-almost automophic for some $1 \leq p < \infty$. Under suitable assumptions the existence of a p-almost automorphic mild solution to $(*)$ is obtained.

1991 Mathematics Subject Classification: Primary 43A60

Keywords and phrases:almost automorphic function, p-almost automorphic function, vanishing mean value, p-locally integrable, p-almost automorphic mild solutions.

1. Introduction

This paper deals with some of the properties of the class of p-almost automorphic functions recently introduced by the author (see [2]) as well as their applications to the abstract differential equations of the form

$$u'(t) = Au(t) + f(t), \tag{1.1}$$

*Communicated by Gaston M. N'Guerekata

†E-mail address: tdiagana@howard.edu

where A is the infinitesimal generator of a c_0-semigroup $(T(t))_{t\geq 0}$ in a Banach space $\mathbb{X}$, and $f : \mathbb{R} \mapsto \mathbb{X}$ is p-almost automophic for some $1 \leq p < \infty$.

We first state and prove some of the properties of the p-almost automorphic functions; next we use them in order to characterize p-almost automorphic mild solutions to (1.1).

If $(\mathbb{X}, \|.\|)$ is a Banach space and if $1 \leq p < \infty$, every (Lebesgue) measurable function f which can be written in a unique fashion as

$$f := h + \phi,$$

where h is almost automorphic and ϕ is p-locally integrable (being an element of $L^p_{loc}(\mathbb{R}, \mathbb{X})$) such that

$$\lim_{T\to+\infty} \frac{1}{2T} \int_{-T}^{T} \|f(\sigma)\|^p \, d\sigma = 0$$

is called p-almost automorphic.

The vector space, as it will be shown, of all (Lebesgue) measurable functions $f : \mathbb{R} \mapsto \mathbb{X}$ satisfying the above-mentioned properties will be called the space of p-almost automorphic functions and denoted by $AA^p(\mathbb{X})$. Many properties of these new spaces as well as their further applications to (1.1) will be discussed. In particular, we shall see that $AA^p(\mathbb{X})$ inherits some of the properties of both $AA(\mathbb{X})$ the Banach space of almost automorphic functions and the classical spaces $L^p_{loc}(\mathbb{R}, \mathbb{X})$ for some $1 \leq p < \infty$.

Let us mention that the present generalization is mainly motivated by abstract differential equations of the form (1.1) where A is a densely defined closed linear operator acting in a Banach space $\mathbb{X}$ and $f : \mathbb{R} \mapsto \mathbb{X}$ is an arbitrary (Lebesgue) measurable function.

Usually, we assume that the function f given in (1.1) is almost automorphic (see [1], [2], [3], [4], [5], [6], [7], and [8])and next one characterizes almost automorphic (mild) solutions of it. However in many interesting applications, the function f is not always almost automorphic. So in such a case how should we deal with the equation (1.1) in terms of almost automorphy? In particular, how should we deal with it when

$$f = g + \phi,$$

where g is almost automorphic and ϕ is a (Lebesgue) measurable function which belongs to a certain class of functions, say $\mathcal{F}(\mathbb{R}, \mathbb{X})$?

Next, how are we going to control the (mild) solution u (if any) in terms of almost automorphy? These are the types of issues we shall address in this paper. In particular, we shall attempt to provide partial answers related to the class of functions $\mathcal{F}(\mathbb{R}, \mathbb{X})$ as well as their further applications to (1.1). Namely, the existence of a p-almost automorphic mild solution to (1.1) will be investigated in *Section 4*.

It should be observed that there is another generalization of almost automophic functions in the literature, the so-called concept of asymptotically almost automorphy. Indeed, if $(\mathbb{X}, \|.\|)$ is a Banach space, a continuous function $f : \mathbb{R}^+ \mapsto \mathbb{X}$ is called asymptotically almost automorphic it it can be expressed as:

$$f(t) = g(t) + \phi(t), \;\; t \in \mathbb{R}^+,$$

where $g : \mathbb{R} \mapsto \mathbb{X}$ is an almost automorphic function and $\phi : \mathbb{R}^+ \mapsto \mathbb{X}$ is a continuous function which vanishes at infinity, ie.,

$$\lim_{s \mapsto +\infty} \|\phi(s)\| = 0.$$

For details on asymptotically almost automorphic functions we refer the reader to [8] and the reference therein. So the notion of p-almost automorphy is welcome to implement the asymptotically almost automorphy.

Throughout the paper $(\mathbb{X}, \|.\|)$ and $L^p_{loc}(\mathbb{X})$ for $1 \leq p < \infty$ stand for a Banach space and the space of (Lebesgue) measurable functions $h : \mathbb{R} \mapsto \mathbb{X}$ which are p-locally integrable.

Let $1 \leq p < \infty$. We define the collection of functions with (asymptotic) vanishing mean value (VMV) of order p by

$$VMV_p(\mathbb{X}) := \{f \in L^p_{loc}(\mathbb{X}) : \lim_{T \to +\infty} \frac{1}{2T} \int_{-T}^{T} \|f(\sigma)\|^p \, d\sigma = 0\}.$$

Lemma 1.1. *The set $VMV_p(\mathbb{X})$ defined above is a vector space.*

Proof. Let $f, g \in VMV_p(\mathbb{X})$ and let $\lambda \in \mathbb{R}$. Since $L^p_{loc}(\mathbb{X})$ is a vector space, it is then clear that both $f + g$ and λf lie in $L^p_{loc}(\mathbb{X})$.

Moreover, using the triangle inequality for the classical norm $\|.\|_p$ of $L^p([-T, T], \mathbb{X})$ for some $T > 0$ it follows that

$$\begin{aligned} 0 &\leq \lim_{T \to +\infty} \frac{1}{2T} \int_{-T}^{T} \|(f+g)(\sigma)\|^p \, d\sigma \\ &\leq [(\lim_{T \to +\infty} \frac{1}{2T} \int_{-T}^{T} \|f(\sigma)\|^p d\sigma)^{1/p} + (\lim_{T \to +\infty} \frac{1}{2T} \int_{-T}^{T} \|g(\sigma)\|^p d\sigma)^{1/p}]^p \\ &= 0. \end{aligned}$$

Therefore, $\lim_{T \to +\infty} \dfrac{1}{2T} \displaystyle\int_{-T}^{T} \|(f+g)(\sigma)\|^p \, d\sigma = 0$, that is, $f + g \in VMV_p(\mathbb{X})$.
In this way, it is easy to see that $\lambda f \in VMV_p(\mathbb{X})$. □

Let us collect some of the classical properties of the almost automophic functions we shall need in the sequel.

Definition 1.2. A continuous function $f : \mathbb{R} \mapsto \mathbb{X}$ is said to be almost automorphic if for every sequence of real numbers (σ_n), there exists a subsequence (s_n) such that

$$g(t) := \lim_{n \mapsto +\infty} f(t + s_n)$$

is well defined for each $t \in \mathbb{R}$, and

$$f(t) = \lim_{n \mapsto +\infty} g(t - s_n),$$

for each $t \in \mathbb{R}$.

As a consequence of the *Definition 1.2*, if $\phi, \psi : \mathbb{R} \mapsto \mathbb{X}$ are $\mathbb{X}$-valued almost automorphic functions and $\lambda \in \mathbb{R}$, then

(i) $\phi + \psi$ is almost automorphic,

(ii) $\lambda\phi$ is almost automorphic,

(iii) $\phi_\lambda(t) := \phi(t + \lambda)$ are almost automorphic,

(iv) $\sup\{\|\phi(t)\|, t \in \mathbb{R}\} < +\infty$,

(v) $R(\phi) := \{\phi(t), t \in \mathbb{R}\}$ is relatively compact.

As stated in (v) the range of an almost automorphic function ϕ is relatively compact on $\mathbb{X}$, therefore it is bounded. Almost automorphic functions constitute a Banach space $AA(\mathbb{X})$ when endowed with the sup norm:

$$\|\phi\|_\infty := \sup_{t\in\mathbb{R}} \|\phi(t)\|.$$

They naturally generalize the concept of *almost periodic functions* as introduced by Bochner in the early sixties. For details on further properties of almost automorphic functions as well as their applications to differential equations we refer readers to ([1, 3, 4, 5], and [8]) and the references therein.

2. p-Almost Automorphic Functions

We are now ready to introduce the space $AA^p(\mathbb{X})$ of p-almost automorphic functions.

Definition 2.1. Let $1 \leq p < \infty$. A (Lebesgue) measurable function $f : \mathbb{R} \mapsto \mathbb{X}$ is said to be p-Almost automorphic if it can be expressed as

$$f = g + \phi, \tag{2.2}$$

where $g \in AA(\mathbb{X})$ and $\phi \in VMV_p(\mathbb{X})$.

The collection of such functions will be denoted by $AA^p(\mathbb{X})$.

The functions g and ϕ given in *Definition 2.1* will be respectively called the *almost automorphic* and the *L^p-perturbation* components of f.

EXAMPLE. Let $1 \leq p < \infty$. An example of a p-almost automorphic function is given by:

$$f(t) = \frac{2 + \exp(-it) + \exp(-i\sqrt{2}t)}{|2 + \exp(-it) + \exp(-i\sqrt{2}t)|} + \frac{e^{-|t|}}{1+t^2}.$$

Clearly, $g(t) = \dfrac{2 + \exp(-it) + \exp(-i\sqrt{2}t)}{|2 + \exp(-it) + \exp(-i\sqrt{2}t)|}$ is almost automorphic while the function $\phi(t) = \dfrac{e^{-|t|}}{1+t^2}$ is in $VMV_p(\mathbb{R})$.

It should be mentioned that the function g given above was introduced by W. A. Veech[10] as an example of an almost automorphic function which is not almost periodic.

The first natural issue we should address concerns the uniqueness of the decomposition in (2.1). Actually, we first show that $AA(\mathbb{X}) \cap VMV_p(\mathbb{X}) = \{0\}$.

We have

Theorem 2.2. *Let f be an almost automorphic function on $\mathbb{R}$. Suppose $f > 0$ and that*

$$\lim_{T \to +\infty} \frac{1}{2T} \int_{-T}^{T} f(t)\,dt = 0. \tag{2.3}$$

Then $f \equiv 0$.

Proof. Suppose $f \not\equiv 0$. Since f is continuous, there exists $\delta > 0$ such that $f(t) \geq \alpha > 1$ for all $t \in (-\delta, \delta)$.

We now show that $f(t) \geq \alpha$ for all $t \in (-\delta, +\infty)$. For that, we use the almost automorphy of f to show that $f(t) \geq \alpha$ for all $t \geq \delta$.

Let $(t_n) = (-n)$ be a sequence of real numbers, where n runs through the set of all natural integers. Now since f is almost automorphic, there exists $(t_{n_k})_{k\in\mathbb{N}} = (-n_k)_{k\in\mathbb{N}}$ a subsequence of (t_n) such that

$$g(t) := \lim_{k \to +\infty} f(t - n_k), \quad \text{for each } \ t \in \mathbb{R}, \tag{2.4}$$

and

$$f(t) = \lim_{k \to +\infty} g(t + n_k), \quad \text{for each } \ t \in \mathbb{R}. \tag{2.5}$$

For all $t \geq \delta$, it is clear that there exists a subsequence $(t_{n_{k_j}})_j$ of $(t_{n_k})_k$ such that $t - n_{k_j} < \delta$ for all $j = 1, 2, 3, \ldots$. Now since $f(t) \geq \alpha > 1$ for all $t \in (-\delta, \delta)$ it easily follows that:

$$f(t - n_{k_j}) \geq \alpha, \quad j = 1, 2, 3, \ldots$$

From (2.3) one obtains:

$$g(t) = \lim_{j \to +\infty} f(t - n_{k_j}) \geq \alpha, \quad \text{for each } \ t \geq \delta,$$

hence $g(t) \geq \alpha$ for all $t \geq \delta$.

Again from the assumption "$f(t) \geq \alpha > 1$ for all $t \in (-\delta, \delta)$" it is clear that $g(t) \geq \alpha$ for all $t \in (-\delta, \delta)$, hence $g(t) \geq \alpha$ for all $t \in (-\delta, +\infty)$.

Using both a similar argument as above and (2.4) it follows that $f(t) \geq \alpha$ for all $t \in (-\delta, +\infty)$.

Now let $T > 0$ (belonging to $(\delta, +\infty)$) we obtain:

$$\begin{aligned} \frac{1}{2T} \int_{-T}^{T} f(t)dt &\geq \frac{1}{2T} \int_{-\delta}^{T} f(t)dt \\ &\geq \frac{\alpha(\delta + T)}{2T} \\ &\geq \frac{\alpha}{2}. \end{aligned}$$

Thus, $\displaystyle \frac{1}{2T} \int_{-T}^{T} f(t)dt \geq \frac{\alpha}{2} \geq \frac{1}{2}$.

Now when T goes to $+\infty$ in the previous inequality it follows that,

$$0 = \lim_{T\to+\infty} \frac{1}{2T} \int_{-T}^{T} f(t)\,dt \geq \frac{1}{2}.$$

The contradiction obtained above completes the proof. □

Using *Theorem 2.2* we obtain:

Corollary 2.3. *If $1 \leq p < \infty$, then $AA(\mathbb{X}) \cap VMV_p(\mathbb{X}) = \{0\}$.*

Proof. We use the fact that if $t \mapsto f(t)$ is almost automorphic, then, so is, $t \mapsto \|f(t)\|^p$. Next, we conclude by *Theorem 2.2* above. □

We are now ready to prove the uniqueness of the decomposition in (2.1). Having in mind that for $1 \leq p < \infty$, $VMV_p(\mathbb{X})$ is a vector one can show the uniqueness of the decomposition in (2.1) as follows: suppose that

$$f = g_1 + \phi_1, \text{ and } f = g_2 + \phi_2,$$

where $g_1, g_2 \in AA(\mathbb{X})$ and $\phi_1, \phi_2 \in VMV_p(\mathbb{X})$.

Now by *Corollary 2.3* it follows that:

$$(g_1 - g_2) = (\phi_2 - \phi_1) \in AA(\mathbb{X}) \cap VMV_p(\mathbb{X}) = \{0\}.$$

Therefore, $g_1 = g_2$ and $\phi_1 = \phi_2$.

In view of the above, it is then clear that $AA(\mathbb{X})$ is a subspace of the space $AA^p(\mathbb{X})$ of p-almost automorphic functions for each $1 \leq p < \infty$.

Combining properties of $AA(\mathbb{X})$ along with those of $L^p_{loc}(\mathbb{X})$ one collects some interesting properties of the space $AA^p(\mathbb{X})$ in the next theorems.

Theorem 2.4. *Let $1 \leq p < +\infty$. If $f, g : \mathbb{R} \mapsto X$ are p-almost automorphic functions and if $\lambda \in \mathbb{R}$, then the following assertions hold true:*

(i) $f + g$ p-almost automorphic;

(ii) $\lambda \,.\, f$ is p-almost automorphic;

(iii) $t \mapsto f(-t)$ is p-almost automorphic;

(iv) $t \mapsto f(t + \lambda)$ is p-almost automorphic.

Proof. The statements (i)-(ii) and (iii) are straightforward. So we only prove (iv). For that, suppose that $f = g + \phi$ is p-almost automorphic, where $g \in AA(\mathbb{X})$ and $\phi \in VMV_p(\mathbb{X})$. Since the function $t \mapsto g(t+\lambda)$ is almost automorphic and that $t \mapsto \phi(t+\lambda)$ is p-locally integrable, to complete the proof we have to show that:

$$\lim_{T\to+\infty} \frac{1}{2T}\int_{-T}^{T} \|\phi(t+\lambda)\|^p\,dt = 0.$$

We have

$$\begin{aligned}
\lim_{T\to+\infty} \frac{1}{2T}\int_{-T}^{T} \|\phi(t+\lambda)\|^p\,dt &= \lim_{T\to+\infty} \frac{1}{2T}\int_{-T+\lambda}^{T+\lambda} \|\phi(t)\|^p\,dt \\
&\le \lim_{T\to+\infty} \frac{1}{2T}\int_{-T-|\lambda|}^{T+|\lambda|} \|\phi(t)\|^p\,dt \\
&= \lim_{T\to+\infty} \frac{1}{2(T+|\lambda|)}\int_{-T-|\lambda|}^{T+|\lambda|} \|\phi(t)\|^p\,dt \\
&= 0.
\end{aligned}$$

This completes the proof. □

Theorem 2.5. *Let $p, q \ge 1$ with $\frac{1}{p} + \frac{1}{q} = 1$ and let $f = g_1 + \phi_1$ and $h = g_2 + \phi_2$, where $g_1, g_2 \in AA(\mathbb{R})$, $\phi_1 \in AA^p(\mathbb{R})$, and $\phi_2 \in AA^q(\mathbb{R})$. Suppose that the product function $g_1.g_2 : t \mapsto g_1(t).g_2(t)$ is almost automorphic on $\mathbb{R}$, then $f.h$ is in $AA^1(\mathbb{R})$.*

Proof. If $f = g_1 + \phi_1$ and $h = g_2 + \phi_2$, where $g_1, g_2 \in AA(\mathbb{R})$, $\phi_1 \in AA^p(\mathbb{R})$, and $\phi_2 \in AA^q(\mathbb{R})$, then $F := f.g$ is given by

$$F = g_1.g_2 + g_2.\phi_2 + \phi_1.g_2 + \phi_1.\phi_2.$$

Set $G = g_1.g_2$ and $\Phi = g_2.\phi_2 + \phi_1.g_2 + \phi_1.\phi_2$. It is then clear that $F = G + \Phi \in AA^1(\mathbb{R})$. By assumption $G \in AA(\mathbb{R})$, this it remains to showing that $\Phi \in VMV_1(\mathbb{R})$. Since every almost automorphic function is bounded for the uniform norm, then $\phi_1.g_2 \in L^p_{loc}(\mathbb{R})$, $g_2.\phi_2 \in L^q_{loc}(\mathbb{R})$, and $\phi_1.\phi_2 \in L^1_{loc}(\mathbb{R})$. Thus $\Phi \in L^1_{loc}(\mathbb{R})$. Moreover,

$$\begin{aligned}
0 &\le \lim_{T\to+\infty} \frac{1}{2T}\int_{-T}^{T} \|\phi_1(t).g_2(t)\|^p\,dt \\
&\le \|g_2\|^p_\infty \cdot \lim_{T\to+\infty} \frac{1}{2T}\int_{-T}^{T} \|\phi_1(t)\|^p\,dt \\
&= 0,
\end{aligned}$$

hence $\phi_1.g_2 \in VMV_p(\mathbb{R})$.

In the same way, $g_2.\phi_2 \in VMV_q(\mathbb{R})$, and $\phi_1.\phi_2 \in VMV_1(\mathbb{R})$. Now since $VMV_s(\mathbb{R}) \subset VMV_r(\mathbb{R})$ whenever $r \le s$ it follows that $\Phi \in VMV_1(\mathbb{R})$. □

3. Further Comments on p-Almost Automorphic Functions

For $1 \leq p < \infty$, define the semi-norm M_p on $AA^p(\mathbb{X})$ by: for all $f \in AA^p(\mathbb{X})$,

$$M_p(f) := (\lim_{T\to\infty} \frac{1}{2T} \int_{-T}^{T} \|f(t)\|^p\, dt)^{\frac{1}{p}}.$$

Using *Theorem 2.2* it can be easily shown that M_p induces a norm on $AA(\mathbb{X})$. In particular if $f = g + \phi$ with $g \in AA(\mathbb{X})$ and $\phi \in VMV_p(\mathbb{X})$,

$$M_p(f) = M_p(g) \leq \|g\|_\infty.$$

Indeed since

$$0 \leq |M_p(f) - M_p(g)| \leq M_p(f-g) = M_p(\phi) = 0,$$

hence $M_p(f) = M_p(g)$.

Now for all $T > 0$ one obtains:

$$\frac{1}{2T} \int_{-T}^{T} \|f(t)\|^p\, dt \leq \|g\|_\infty^p,$$

hence $M_p(f) \leq \|g\|_\infty$.

4. Existence of p-almost Automorphic Mild Solutions

We now investigate upon the existence of a p-almost automorphic mild solution $u : \mathbb{R} \mapsto \mathbb{X}$ to (1.1). For that, we require the following assumptions:

(H.1) A is the infinitesimal generator of a c_0-semigroup $(T(t))_{t\geq 0}$ on $\mathbb{X}$ such that

$$\int_0^{+\infty} \|T(\sigma)u\|\, d\sigma < +\infty, \quad \forall u \in \mathbb{X},$$

(H.2) $f \in AA^1(\mathbb{X})$, ie., $f = g + \phi$, where $g \in AA(\mathbb{X})$ and $\phi \in VMV_1(\mathbb{X})$.

we have

Theorem 4.1. *Under assumptions (H.1)-(H.2), suppose that ϕ the L^1-component of f is in $L^1(\mathbb{R}, \mathbb{X})$. Then (1.1) has a 1-almost automorphic mild solution.*

The proof of the *Theorem 4.1* requires the following lemma (for details we refer the reader to[9, Theorem 4.1, p. 116]):

Lemma 4.2. *Let A be the infinitesimal generator of a c_0-semigroup $(T(t))_{t\geq 0}$ in a Banach space $\mathbb{X}$. If for some $1 \leq q < \infty$*

$$\int_0^{+\infty} \|T(\sigma)u\|^q \, d\sigma < +\infty, \ , \ \forall t \geq 0,$$

$\forall u \in \mathbb{X}$, then there exist constants $M \geq 1$ and $\varpi > 0$ such that

$$\|T(t)\| \leq Ke^{-\varpi t} \qquad (t \geq 0).$$

We are now ready to prove *Theorem 4.1*.

Proof. (*Theorem 4.1*). Let u be a mild solution to (1.1). It is well-known that such a mild solution can be expressed as:

$$u(t) = T(t-a)u(a) + \int_a^t T(t-s)f(s)ds,$$

for all $a \in \mathbb{R}$ and $t \geq a$.

Now according to (H.2) one can write $f = g + \phi$. Thus the mild solution u can be expressed as $u = F + \Phi$, where

$$F(t) := T(t-a)u(a) + \int_a^t T(t-s)g(s)ds,$$

and

$$\Phi(t) := \int_a^t T(t-s)\phi(s)ds.$$

We shall show that $F \in AA(\mathbb{X})$ and $\Phi \in VMV_1(\mathbb{X})$. Indeed, according to assumption (H.1) and *Lemma 4.2* above, there are constants $K \geq 1$ and $\omega > 0$ such that

$$\|T(t)\| \leq Ke^{-\omega t}, \ \forall t \geq 0. \tag{4.6}$$

Using (4.1) and following along the same lines as the proof of [7, Theorem 3.1] it follows that $t \mapsto F(t)$ is almost automorphic.

It remains to show that $\Phi \in VMV_1(\mathbb{X})$. Now using (4.1) and the fact that $\phi \in L^1(\mathbb{X})$ it easily follows that Φ is bounded, in particular $\Phi \in L^1_{loc}(\mathbb{R}, \mathbb{X})$.

To complete the proof we have to show that:

$$\lim_{T\to+\infty} \frac{1}{2T} \int_{-T}^{T} \|\Phi(t)\| \, dt = 0.$$

Again, from (4.1) it is clear that

$$0 \leq \lim_{T\to+\infty} \frac{1}{2T} \int_{-T}^{T} \|\Phi(s)\| \, ds \leq I + J,$$

where

$$I := \lim_{T\to+\infty} \frac{K}{2T} \int_{-T}^{T} \|\phi(s)\| \, ds \int_{-T}^{s} e^{-\omega(s-\sigma)} \, d\sigma,$$

and

$$J := \lim_{T\to+\infty} \frac{K}{2T} \int_{-T}^{T} ds \int_{-\infty}^{-T} e^{-\omega(s-\sigma)} \|\phi(\sigma)\| \, d\sigma.$$

Now

$$\begin{aligned} I &= \lim_{T\to+\infty} \frac{K}{2T} \int_{-T}^{T} \|\phi(s)\| \, ds \int_{-T}^{s} e^{-\omega(s-\sigma)} \, d\sigma \\ &= \lim_{T\to+\infty} \frac{K}{2T} \int_{-T}^{T} \|\phi(s)\| \, \frac{1}{\omega}(1 - e^{-\omega(s+T)}) \, ds. \end{aligned}$$

Since $-T \leq s \leq T$ and $\omega > 0$ it is clear that $\dfrac{1}{\omega}(1 - e^{-\omega(s+T)}) \leq \dfrac{1}{\omega}$. It follows that

$$\begin{aligned} I &= \frac{K}{\omega} \cdot \lim_{T\to+\infty} \frac{1}{2R} \int_{-T}^{T} \|\phi(s)\| \, ds \\ &= 0. \end{aligned}$$

It remains to show that $J = 0$. Actually, this requires much more computations. We have

$$\begin{aligned} J &= \lim_{T\to+\infty} \frac{K}{2T} \int_{-T}^{T} ds \int_{-\infty}^{-T} e^{-\omega(s-\sigma)} \|\phi(\sigma)\| \, d\sigma. \\ &= \lim_{T\to+\infty} \frac{K}{2T} \int_{-\infty}^{-T} \|\phi(\sigma)\| \, d\sigma \int_{-T}^{T} e^{-\omega(s-\sigma)} ds \\ &= \lim_{T\to+\infty} \frac{K}{2T} \int_{-\infty}^{-T} \|\phi(\sigma)\| \, \frac{1}{\omega} \, (e^{-\omega(-T-\sigma)} - e^{-\omega(T-\sigma)}) \, d\sigma \\ &= \lim_{T\to+\infty} \frac{K}{2T\omega} \int_{-\infty}^{-T} \|\phi(\sigma)\| \, (e^{\omega(T+\sigma)} - e^{-\omega(T-\sigma)}) \, d\sigma \\ &= \lim_{T\to+\infty} \frac{K}{2T\omega} \, [\int_{-\infty}^{-T} \|\phi(\sigma)\| e^{\omega(T+\sigma)} \, d\sigma - \int_{-\infty}^{-T} \|\phi(\sigma)\| e^{-\omega(T-\sigma)} \, d\sigma] \end{aligned}$$

Now note that $T + \sigma \leq 0$ for all $\sigma \leq -T$. Since $\phi \in L^1(\mathbb{R}, \mathbb{X})$ it is then clear that

$$\begin{aligned} \int_{-\infty}^{-T} \|\phi(\sigma)\| e^{\omega(T+\sigma)} \, d\sigma &\leq \int_{-\infty}^{-T} \|\phi(\sigma)\| \, d\sigma \\ &\longrightarrow 0, \quad \text{as } T \mapsto +\infty. \end{aligned}$$

Hence, $\lim_{T\to+\infty} \dfrac{K}{2T\omega} \cdot \displaystyle\int_{-\infty}^{-T} \|\phi(\sigma)\| e^{\omega(T+\sigma)} \, d\sigma = 0$.

In the same way one has: $T - \sigma \geq 2T > 0$ for all $\sigma \leq -T$. Thus using similar arguments as above it follows that

$$\begin{aligned} \int_{-\infty}^{-T} \|\phi(\sigma)\| e^{-\omega(T-\sigma)} \, d\sigma &\leq \int_{-\infty}^{-T} \|\phi(\sigma)\| \, d\sigma \\ &\mapsto 0, \quad \text{as } T \mapsto +\infty. \end{aligned}$$

Here again, $\lim_{T\to+\infty} \frac{K}{2T\omega} \cdot \int_{-\infty}^{-T} \|\phi(\sigma)\| e^{-\omega(T-\sigma)}\, d\sigma = 0.$

Therefore $J = 0$.

In view of the above it follows that $\Phi \in VMV_1(\mathbb{X})$, hence the mild solution $u = F + \Phi$ belongs to $AA^1(\mathbb{X})$. Actually, F is the almost automorphic component of u while Φ is its L^1-component. □

REMARK. For $1 \leq p < \infty$, consider the class all (Lebesgue) measurable functions $F : \mathbb{R} \times \mathbb{X} \mapsto \mathbb{X}$. As in *Definition 2.1* one can define $AA^p(\mathbb{R} \times \mathbb{X})$ as follows: $F : \mathbb{R} \times \mathbb{X} \mapsto \mathbb{X}$ will be called p-almost automorphic (being an element of $AA^p(\mathbb{R} \times \mathbb{X})$) if it can be decomposed as

$$F = G + \Phi,$$

where $G : \mathbb{R} \times \mathbb{X} \mapsto \mathbb{X}$, $(t, u) \mapsto F(t, u)$ is almost automorphic in $t \in \mathbb{R}$ for each $u \in \mathbb{X}$, and $\Phi \in VMV(\mathbb{R} \times \mathbb{X})$ with $VMV_p(\mathbb{R} \times \mathbb{X})$ being the collection of all functions $t \mapsto \Psi(t, u) \in L^p_{loc}(\mathbb{X})$ such that

$$\lim_{T\to+\infty} \frac{1}{2T} \int_{-T}^{T} ||\Psi(t, u)||^p\, dt = 0,$$

uniformly in $u \in \mathbb{X}$.

REMARK. The previous remark should enable us to study p-almost automorphic solutions to semilinear differential equations of the form

$$\frac{du}{dt} = Au(t) + F(t, u(t)), \tag{4.7}$$

where A is a densely defined closed operator on $\mathbb{X}$, and $F \in AA^p(\mathbb{R} \times \mathbb{X})$.

References

[1] T. Diagana, Some Remarks on Some Second-Order Hyperbolic Differential Equations. *Semigroup Forum*, Vol. **68** (2004) 357-364.

[2] T. Diagana, *A Generalization of Almost Automorphic Functions I*, Preprint (2004).

[3] T. Diagana, G. M. N'Guérékata and N. V. Minh, Almost Automorphic Solutions of Evolution Equations, *Proc. Amer. Math. Soc.*, **132** (2004), no. 11, pp. 3289–3298.

[4] T. Diagana and G. M. N'Guerekata, On Some Perturbations of Some Abstract Differential Equations. *Comment. Math.* **XLIII** (2) (2003), pp 201–206.

[5] T. Diagana and G. M. N'Guerekata, Some Extension of the Bohr-Neugebauer-N'Guerekata Theorem. *J. of Anal. Appl.* Vol. **2** (2004), no.1, pp. 1–10.

[6] J. A. Goldstein and G. M. N'Guérékata, Almost Automorphic Solutions of Semilinear Evolution Equations, *Proceedings of the American Mathematical Society* (in press).

[7] G. M. N'Guerekata, Existence and Uniqueness of Almost Automorphic Mild Solutions to Some Semilinear Abstract Differential Equations, *Semigroup Forum*, Vol. **69**(2004) 80-86.

[8] G. M. N'Guerekata, *Almost Automorphic Functions and Almost Periodic Functions in Abstract Spaces*, Kluwer Academic / Plenum Publishers, New York-London-Moscow, 2001.

[9] A. Pazy, *Semigroups of linear operators and applications to partial differential equations*. Springer-Verlag-Berlin-New York, 1983.

[10] W. A. Veech, Almost Automorphic Functions. *Proc. Nat. Acad. Sci. U.S.A.* Vol. **49**(1963), 462–464.

[11] W. A. Veech, Almost Automorphic Functions on Groups. *Amer. J. Math.* Vol. **87**(1965), 719–751.

[12] S. Zaidman, Almost Automorphic Solutions of Some Abstract Evolution Equations. II. *Istit. Lombardo. Accad. Sci. Lett. Rend. A* **111**(1977), no. 2, 260–272.

[13] S. Zaidman, Almost Automorphic Solutions of Some Abstract Evolution equations. *Istit. Lombardo. Accad. Sci. Lett. Rend. A* **110**(1976), no. 2, 578–588.

[14] S. Zaidman, Almost-Periodic Functions in Abstract Spaces, *Research Notes in Mathematics* **126**, Pitman Advanced Publishing Program, Boston. London. Melbourne, 1985.

[15] M. Zaki, Almost Automorphic Solutions of Certain Abstract Differential Equations. *Ann. Mat. Pura Appl*. Vol. **101** (1974), no. 4, 91–114.

[16] M. Zaki, Almost Automorphic Integrals of Almost Automorphic functions. *Canad. Math. Bull.* Vol. **15** (1972), 433–436.

[17] M. Zaki, Almost Automorphic Solutions of Non-Homogeneous Heat Equation. *Ann. Univ. Ferrara Sez*. VII (N.S.) Vol. **16** (1971), 149–156.

[18] M. Zima, A certain fixed point theorem and its application to integral- functional equations, *Bull. Austral. Math. Soc*. **46**(2)(1992), 179-186.

In: Focus on Evolution Equations
Editor: Gaston M. N'Guerekata
ISBN: 978-1-60021-342-7

Chapter 4

LYAPUNOV FUNCTION FOR A BIDIMENSIONAL SYSTEM-MODEL IN THE GLUCOSE HOMEOSTASY AND CLINICAL INTERPRETATIONS*

Alexandru Bica†
University of Oradea, Department of Mathematics,
Oradea, Romania

Abstract

In this paper we study the model proposed in [9], which describe the glucose homeostasy. The first approximation analysis and a Lyapunov function are developed to obtain the phase portrait. Each result has clinical interpretation which is presented in terms of the dynamical equilibrium between the glucose and hormonal levels.

2000 Mathematics Subject Classification: 92A09.

Keywords and phrases: nonlinear autonomous differential system, linear approximation, Lyapunov function.

1. Introduction

In this paper, we continue the work with the bidimensional model (1) from [9], where $G(t)$ is the glucose level in blood at the moment t, $I(t)$ is the insulinemia at the moment t, $H(t)$ is an average of the hyperglycaemiant hormonal levels in blood at the same moment, G_0, I_0, H_0 are the equilibrium homeostasic levels measured in the morning after fast overnight. Here, we denote, $x(t) = G(t) - G_0$, and $y(t) = I(t) - I_0 - (H(t) - H_0)$ The model is governed by the following nonlinear autonomous differential system :

$$\begin{cases} x' = -a\dfrac{xy}{x + G_0} - bx - my = F(x,y) \\ y' = cx - dy = G(x,y) \end{cases} \tag{1}$$

*Communicated by Sorin G. Gal

†E-mail addresses: smbica@yahoo.com, abica@uoradea.ro

where $a, b, c, d, m > 0$ are constants. Here, $F, G : (-G_0, \infty) \times \mathbb{R} \to \mathbb{R}$ have continuous partial derivatives.

In the study of glycaemic homeostasy there are many papers from 1965 to present. All these models are dedicated to control the diabete disease. Most of these models use the glucose tolerance test (see [1], [3], [4], [5], [7], [8], [10], [11], [12], [13], [14], [17], [18], [20]). Some of the glucose-insulin models become classical in time, such as the models from [1], [2], [11], and [14]. The recent models are governed by delay differential equations; for instance, the model from [15] uses the results from [6], [19], [21], [22] and generalizes the De Gaetano-Arino model (see [14]), explains the glucose tolerance test and presents results of local and global asymptotic stability of the unique equilibrium point. Another recent model was developed by the Lithuanian team from Klaipeda and Vilnius: Basov, Meilũnas, Švitra, where the model is obtained in [3], using an idea of prey-predator interaction between the glucose and insulin, with the insulin containing a delay component. For this model it is proved the existence and uniqueness of a periodic solution , which explains the experimental data from [7] and [23] corresponding to a diabetic and a healthy person, respectively. If we see with attention the data from [23] presented in [3] (in the figure 3), we observe that the values of the blood-glucose in the time interval 6-20 hours (on a day) are very closely to an horizontally line (exception is the value at 13'30 hour which presents a normal increasing value corresponding to lunch) and for this reason we don't believe in the periodic evolution of blood-glucose levels during a day. We think that periodic is only the circadian rhythm. For this reason we try to study the equilibrium solutions of the system (1) and the perturbations from these points, to describe the blood-glucose level evolution. Moreover, we intend to obtain some information from the above (1) model (which is valid for a healthy man) for the prevention of metabolic disorders and the stop at the initial stage of the diabete disease. The clinical interpretations will be used for this aim of the disease prevention.

In the system (1), the nonlinear term $-\dfrac{axy}{x+G_0}$ represents the interaction glucose-insulin (together with the hiperglycaemiant hormones) in the reversible transformation process glucose-glycogen from liver. Therefore, the constant a is the efficiency of the liver. This nonlinear term describe a prey- predator interaction with the insulin as predator for glucose (because in liver, the glucose is transformed to glycogen by insulin and the glycogen is transformed to glucose by the glucagon consumption : the glucose is predator for glucagon and for other hiperglycaemiant hormones). This interaction is of Michaelis-Menten type, because the tendency of increase or decrease of the blood-glucose level is braked by the actual glycaemy. Without the nonlinear term, the system (1) becomes the same as the linear system from [1], but the variable y has another significance than in [1].

Since the glycaemic homeostasy mechanism involve negative feedback, the coefficients b and d describe the efficiency of the negative self-feedback control of glucose and hormones, respectively. The coefficient m represent the efficiency of hormonal action in tissue in the sense to increase or decrease the glycaemy (for instance, this coefficient describe the insulin catalization to penetrate the cellular membrane for the uptake of the glucose in the tissular cells). The coefficient c is the efficiency of the glucorecievers and of the glucose action to modify the hormonal level.

Firstly, we study the local stability of the equilibrium solutions of the system (1) using the first approximation method and we determine the attraction basin of the origin and the phase portrait in the neighborhood of the equilibrium points. Afterwards, we construct the Lyapunov function and establish the domain of global stability. Also, we prove that usually there is no periodic solutions for the system (1). Finally we compare our results with other known in the literature.

2. Local Stability of the Equilibrium Points

The equilibrium solutions of the system (1) are the solutions of the nonlinear algebraic system

$$\begin{cases} F(x,y)=0 \\ G(x,y)=0 \end{cases}$$

denoted by $P_1(0,0)$ and $P_2(x_2,y_2)$, where

$$x_2 = \frac{-G_0(bd+mc)}{ac+bd+mc} < 0$$

and

$$y_2 = \frac{-cG_0(bd+mc)}{d(ac+bd+mc)} < 0.$$

Theorem 1. *For any positive values of the coefficients* a,b,c,d,m *the solution* $(0,0)$ *of the system (1) is uniformly asymptotic stable and the equilibrium point* $P_2(x_2,y_2)$ *is a saddle point. If* $(b-d)^2 < 4mc$ *then the solution* $(0,0)$ *is asymptotic stable focus, and if* $(b-d)^2 \geq 4mc$ *then the solution* $(0,0)$ *is asymptotic stable node.*

Proof. The eigenvalue equation in the neighborhood of the origin is

$$\lambda^2 + (b+d)\lambda + bd + mc = 0. \tag{2}$$

Since $b+d>0$ and $bd+mc>0$, we infer that $\operatorname{Re}\lambda_{1,2} < 0$, $\forall b,c,d,m>0$. The discriminant of the equation (2) is $\Delta=(b-d)^2-4mc$ and therefore $\lambda_{1,2}\in\mathbb{C}\setminus\mathbb{R} \iff (b-d)^2 < 4mc$. The eigenvalue equation in the neighborhood of the point P_2 is

$$\lambda^2 + \lambda[d - \frac{(bd+mc)^2+amc^2}{acd}] - bd - \frac{(bd+mc)^2+amc^2}{ac} = 0.$$

Since, $\lambda_1\cdot\lambda_2 = -bd - \frac{(bd+mc)^2+amc^2}{ac} < 0$, $\forall a,b,c,d,m>0$, we infer that the point $P_2(x_2,y_2)$ is a saddle point. □

Remark 1. *The stable and the unstable manifold of the saddle point* P_2 *are smooth curves and the unstable manifold realizes the connection between the equilibrium points. The global stable manifold is the separatrix of the attraction basin of the origin. Using the method from [16], page 52, we compute the slope of each manifold through* P_2*. Since the matrix of the linear approximation in the neighborhood of* P_2 *is,*

$$J(P_2) = \begin{pmatrix} \frac{(bd+mc)^2+amc^2}{acd} & \frac{bd}{c} \\ c & -d \end{pmatrix}$$

the corresponding linear system can take the form of a homogeneous Euler's type differential equation, where $x = \xi$ and $y = \eta$,

$$\frac{d\eta}{d\xi} = \frac{c\xi - d\eta}{\frac{(bd+mc)^2+amc^2}{acd} \cdot \xi + \frac{bd}{c} \cdot \eta}. \tag{3}$$

We look for lines through the origin as solutions of the equation (3), that is $\eta = k\xi$, with $k \in \mathbb{R}^*$. Then the equation for k is,

$$k = \frac{c - kd}{\frac{(bd+mc)^2+amc^2}{acd} + \frac{bd}{c} \cdot k}$$

that is,

$$k^2 + \frac{(bd+mc)^2 + ac(mc+d^2)}{abd^2} \cdot k - \frac{c^2}{bd} = 0. \tag{4}$$

Then, the roots are $k_1 < 0$ and $k_2 > 0$, $\forall a, b, c, d, m > 0$, where,

$$k_{1,2} = \frac{-[(bd+mc)^2 + ac(mc+d^2)] \pm ([(bd+mc)^2 + ac(mc+d^2)]^2 + 4adc^2)^{\frac{1}{2}}}{2abd^2}.$$

Here, k_1 is the slope at P_2 of the stable manifold, and k_2 is the slope in P_2 of the unstable manifold.

Remark 2. *The linear approximation in the neighborhood of the origin is the system*

$$\begin{cases} x' = -bx - my \\ y' = cx - dy \end{cases}.$$

the same as in [1]. If we suppose that $(b-d)^2 < 4mc$ and denote $\alpha = \frac{1}{2}(b+d)$, $\omega_0^2 = bd + mc$, $\omega^2 = \omega_0^2 - \alpha^2$ then we obtain the solutions of this linear system, $x(t)$, $y(t)$, as linear combinations of the damping oscillations of the form : $e^{-\alpha t}\cos\omega t$, $e^{-dt}\cos\omega t$ and $e^{-\alpha t}\sin\omega t$.

Remark 3. *If we denote $\mu = bd$, then we can study the dependence on μ of the distance between x_1 and x_2, $d(x_1, x_2) = D(\mu)$. So,*

$$D(\mu) = \left| \frac{-G_0(bd+mc)}{ac+bd+mc} \right| = \frac{G_0(\mu+mc)}{\mu+ac+mc}$$

and

$$D'(\mu) = \frac{acG_0}{(\mu+ac+mc)^2} > 0, \ \forall \mu > 0,$$

which means that $D(\mu)$ is increasing. We infer that the increasing of the negative self-feedback coefficients b and d lead to the increasing of the distance between the equilibrium points.

Remark 4. *The clinical interpretation of Theorem 1 and of Remarks 2 and 3 are as follows: Since the saddle point P_2 can be viewed as a critical point (from the medical point of view) and it is a frontier glucose level which separate the homeostasy by the chaotic (dangerous) hipoglycaemia, we can say that a great distance $D(\mu)$ represent a chance of the impossibility to reach this critical point and so, the glycaemic values will be only closely to the origin. A person which have this property have small chances to reach (in the future) the diabet disease because his stability of the glycaemic homeostasic value is strong when b and d have great values (the damping oscillations from the above remark are strongly damped and therefore any perturbation from origin will fastly arrive at the origin). We can conclude that the distance between the equilibrium points is a good characterization for a healthy person for present and for the future. On the other hand, a person for which this distance is small have low values of b and d and consequently, the glycaemy can arrive frequently in a neighborhood of P_2 requiring often feeding. Such persons, healthy in the youth period, have the risk to became diabetic after 45-50 years old, because the stability of the glycaemic homeostasic value is more weak (have predisposition to hipoglycaemya in the youth period and to hiperglicaemya at the begging of the old age). Therefore, the persons which require, in the youth adult period, frequent feeding, must to expect to hiperglycaemya after 45 years and so, must to prevent this possibility.*

We see that the stable manifold of P_2 is the frontier of the attraction basin of the origin, but we can determine this frontier only numerically. To determine analytically greater subset of this attraction basin we will study the global stability of the solution $(0,0)$ of the system (1).

3. Main Result

Let $x^* = \frac{-mG_0}{a+m}$. We see that

$$\frac{ax}{x+G_0} + m > 0,\ \forall x > x^*$$

and so, we can define the Lyapunov function on $(x^*, +\infty) \times \mathbb{R}$.

Theorem 2. *If $\beta > 0$ and $\alpha : (x^*, +\infty) \longrightarrow \mathbb{R}$ is a continuous function such that $\alpha(0) = 0$, $\alpha(x) > 0$, $\forall x \in (x^*, +\infty) \setminus \{0\}$ and $\alpha(x) \leq 8\beta bdmx(\frac{ax}{x+G_0} + m)^{-4}$, $\forall x \geq 0$, then the function $V : (x^*, +\infty) \times \mathbb{R} \longrightarrow \mathbb{R}$,*

$$V(x,y) = \begin{cases} \int_x^0 \alpha(s)(\frac{as}{s+G_0} + m)ds + \beta y^2, & x \in (x^*, 0] \\ \int_0^x \alpha(s)(\frac{as}{s+G_0} + m)ds + \beta y^2, & x \in [0, \infty) \end{cases} \tag{5}$$

is a Lyapunov function of the autonomous system (1), on the set

$$\{(x,y) \in \mathbb{R}^2 : x > \frac{-mG_0}{a+m}\}.$$

Proof. We can see that

$$\lim_{x\to 0, x<0} V(x,y) = \lim_{x\to 0, x<0} V(x,y) = V(0,y) = \beta y^2, \forall y \in \mathbb{R},$$

that is V is continuous on $(x^*, +\infty) \longrightarrow \mathbb{R}$. Moreover,

$$\frac{\partial V}{\partial x}(x,y) = \begin{cases} -\alpha(x)(\frac{ax}{x+G_0}+m), & x \in (x^*, 0] \\ \alpha(x)(\frac{ax}{x+G_0}+m), & x \in [0, \infty) \end{cases}$$

$$\frac{\partial V}{\partial y}(x,y) = 2\beta y, \quad \forall (x,y) \in (x^*, +\infty) \times \mathbb{R}$$

and

$$\lim_{x\to 0, x<0} \frac{\partial V}{\partial x}(x,y) = \lim_{x\to 0, x<0} \frac{\partial V}{\partial x}(x,y) = 0, \forall y \in \mathbb{R},$$

since $\alpha(0) = 0$. Then $V \in C^1((x^*, +\infty) \times \mathbb{R})$. Because $\alpha(x) > 0$, $\frac{ax}{x+G_0} + m > 0$, $\forall x > x^*, x \neq 0$, we infer that $V(x,y) > 0$, $\forall (x,y) \in (x^*, +\infty) \times \mathbb{R} \setminus \{(0,0)\}$ and see that $V(0,0) = 0$. Also, we see that

$$\lim_{\|(x,y)\|\to\infty} V(x,y) = +\infty.$$

We study now, the sign of the time derivative, $\frac{d}{dt}V(x(t), y(t))$. For $(x,y) \in (x^*, 0] \times \mathbb{R}$, we have,

$$\frac{d}{dt}V(x(t),y(t)) = \frac{\partial V}{\partial x}(x,y)\cdot x'(t) + \frac{\partial V}{\partial y}(x,y)\cdot y'(t) =$$

$$= -\alpha(x)(\frac{ax}{x+G_0}+m)\cdot(-a\frac{xy}{x+G_0} - bx - my) + 2\beta y(cx - dy) =$$

$$= \alpha(x)(\frac{ax}{x+G_0}+m)^2\cdot y + bx\alpha(x)(\frac{ax}{x+G_0}+m) + 2\beta cxy - 2d\beta y^2 =$$

$$= -2d\beta y^2 + y\cdot[2\beta cx + \alpha(x)(\frac{ax}{x+G_0}+m)^2] + bx\alpha(x)(\frac{ax}{x+G_0}+m) = R_2(y).$$

We see that $-2d\beta y^2 < 0$ and $bx\alpha(x)(\frac{ax}{x+G_0}+m) \leq 0$, $\forall y \in \mathbb{R}$, $\forall x \in (x^*, 0]$. We will determine the function α on the interval $(x^*, 0]$, such that

$$2\beta cx + \alpha(x)(\frac{ax}{x+G_0}+m)^2 = 0,$$

that is,

$$\alpha(x) = -2\beta cx(\frac{ax}{x+G_0}+m)^{-2} > 0, \ \forall x \in (x^*, 0],$$

with $\alpha(0) = 0$, $\lim_{x\to 0, x<0} \alpha(x) = 0$ and $\lim_{x\to x^*, x>x^*} \alpha(x) = +\infty$. Then,

$$\frac{d}{dt}V(x(t),y(t)) = -2d\beta y^2 + bx\alpha(x)(\frac{ax}{x+G_0}+m) < 0, \forall (x,y) \in (x^*, 0] \times \mathbb{R}. \quad (6)$$

For $(x,y) \in [0,\infty) \times \mathbb{R}$, we have,

$$\frac{d}{dt}V(x(t),y(t)) = \frac{\partial V}{\partial x}(x,y)\cdot x'(t) + \frac{\partial V}{\partial y}(x,y)\cdot y'(t) =$$

$$= \alpha(x)(\frac{ax}{x+G_0} + m) \cdot (-a\frac{xy}{x+G_0} - bx - my) + 2\beta y(cx - dy) =$$

$$= -2d\beta y^2 - y \cdot [\alpha(x)(\frac{ax}{x+G_0} + m)^2 - 2\beta cx] - bmx\alpha(x) -$$

$$-\frac{ab\alpha(x)}{x+G_0} \cdot x^2 = Q_2(y) - \frac{ab\alpha(x)}{x+G_0} \cdot x^2.$$

The discriminant of $Q_2(y)$ is, for $x \geq 0$,

$$\Delta(x) = [\alpha(x)(\frac{ax}{x+G_0} + m)^2 - 2\beta cx]^2 - 8\beta bdmx\alpha(x).$$

For any $x \geq 0$, we have,

$$\alpha(x)(\frac{ax}{x+G_0} + m)^2 - 2\beta cx \leq \alpha(x)(\frac{ax}{x+G_0} + m)^2,$$

and if $\alpha(x) \leq 8\beta bdmx(\frac{ax}{x+G_0} + m)^{-4}, \forall x \geq 0$, then

$$[\alpha(x)(\frac{ax}{x+G_0} + m)^2 - 2\beta cx]^2 \leq 8\beta bdmx\alpha(x), \forall x \geq 0.$$

Consequently, $\Delta(x) \leq 0, \forall x \geq 0$ and so, $Q_2(y) \leq 0, \forall x \geq 0, \forall y \in \mathbb{R}$. Since, $\frac{ab\alpha(x)}{x+G_0} \cdot x^2 > 0, \forall x > 0$, we deduce that

$$\frac{d}{dt}V(x(t), y(t)) < 0, \forall (x, y) \in (0, \infty) \times \mathbb{R}. \quad (7)$$

In the case $x \in [0, \infty)$ we can consider a function $\alpha(x) \leq 8\beta bdmx(\frac{ax}{x+G_0} + m)^{-4}, \forall x \geq 0$. Since,

$$\frac{1}{(a+m)^4} \leq (\frac{ax}{x+G_0} + m)^{-4} \leq \frac{1}{m^4}, \forall x \geq 0,$$

we can take $\alpha(x) = \frac{Ax}{(a+m)^4}, \forall x \in [0, \infty)$, where $A \in (0, \infty)$, $A < 8\beta bdm$. Then, for $\alpha : (x^*, +\infty) \longrightarrow \mathbb{R}$, we have the expression,

$$\alpha(x) = \begin{cases} -2\beta cx(\frac{ax}{x+G_0} + m)^{-2}, & x \in (x^*, 0] \\ \frac{Ax}{(a+m)^4}, & x \in [0, \infty), \end{cases} \quad (8)$$

where $A \in (0, 8\beta bdm)$. From (6) and (7) we infer that $\frac{d}{dt}V(x(t), y(t)) < 0, \forall (x, y) \in (x^*, \infty) \times \mathbb{R}$, for the function α defined in (8). We conclude that the function V defined by (5) is a Lyapunov function for the system (1) on the set

$$\{(x, y) \in \mathbb{R}^2 : x > \frac{-mG_0}{a+m}\}.$$

According to the La Salle invariance principle, we infer that the solution $(0, 0)$ of the system (1) is globally asymptotic stable on the set $(-\frac{mG_0}{a+m}, +\infty) \times \mathbb{R}$. □

To establish the phase portrait we use the nullclines,

$$\begin{cases} F(x,y) = -a\dfrac{xy}{x+G_0} - bx - my = 0 \\ G(x,y) = cx - dy = 0 \end{cases}$$

and their intersection is the points $P_1(0,0)$ and $P_2(x_2,y_2)$. These two nullclines are the line $y = \frac{c}{d}x$ and the plane curve with two branches,

$$y = \frac{-bx(x+G_0)}{(a+m)x + mG_0} \tag{9}$$

which has the same vertical asymptote, $x = -\frac{mG_0}{a+m}$. Since $\lim\limits_{bd\to 0} \frac{-G_0(bd+mc)}{ac+bd+mc} = -\frac{mG_0}{a+m}$, we infer that the point $P_2(x_2,y_2)$ lies on the branch of the curve (9) which is on the left side of the asymptote $x = -\frac{mG_0}{a+m}$ and the origin lies on the branch which is on the right side of this asymptote contained on the set $(-\frac{mG_0}{a+m}, +\infty) \times \mathbb{R}$, of global stability.

4. The Attraction Basin and the Nonexistence of the Periodic Solutions

The two nullclines determine the arrows of the field lines of the system (1) and divide the set $(-G_0,\infty) \times \mathbb{R}$ in six regions. The set $(-\frac{mG_0}{a+m}, +\infty) \times \mathbb{R}$ is the first region (region (I)), the second set is $(-G_0, -\frac{mG_0}{a+m}) \times [0,\infty)$ (region (II)). The region (III) is the set,

$$\{(x,y) \in \mathbb{R}^2 : -G_0 < x < -\frac{mG_0}{a+m},\ y < 0,\ y > \frac{c}{d}x,\ y > \frac{-bx(x+G_0)}{(a+m)x+mG_0}\},$$

the region (IV) is the set,

$$\{(x,y) \in \mathbb{R}^2 : -G_0 < x < -\frac{mG_0}{a+m},\ \frac{-bx(x+G_0)}{(a+m)x+mG_0} < y < \frac{c}{d}x\},$$

the region (V) is the set,

$$\{(x,y) \in \mathbb{R}^2 : -G_0 < x < -\frac{mG_0}{a+m},\ \frac{c}{d}x < y < \frac{-bx(x+G_0)}{(a+m)x+mG_0}\}$$

and the region (VI) is the set,

$$\{(x,y) \in \mathbb{R}^2 : -G_0 < x < -\frac{mG_0}{a+m},\ y < \frac{c}{d}x,\ y < \frac{-bx(x+G_0)}{(a+m)x+mG_0}\}.$$

In the region (I) is the set of global stability. In the region (II) we have $x < 0, y > 0, x' > 0$ and $y' < 0$, and then

$$\frac{d}{dt}[(x^2+y^2)(t)] = 2x(t)\cdot x'(t) + 2y(t)\cdot y'(t) < 0, \forall t \geq 0,$$

that is $r(t) = [(x^2+y^2)(t)]^{\frac{1}{2}}$ decrease in this region. This means that any orbit which starts from the region (II) will arrive at the origin. In the region (IV) we have $x < 0, y < 0, x' > 0, y' > 0$ and then $\frac{d}{dt}[(x^2+y^2)(t)] < 0, \forall t \geq 0$. Then in the region (IV) the $\omega-$limit set of each point is the origin. In the region (III) we have, $y < 0, y' < 0, x' > 0, x < x^*$ and therefore $2y(t) \cdot y'(t) > 0$ and

$$2(x(t) - x^*) \cdot x'(t) = \frac{d}{dt}[(x(t) - x^*)^2] < 0, \quad \forall t \geq 0,$$

So, any orbit which starts from this region is conducted to the region (I), or to the region (IV).

In the region (V) we have $y < 0, y' < 0, x' < 0, x < x^*$ and in the region (VI), $y < 0, y' > 0, x' > 0, x < x^*$. Then each orbit which starts from this region move away on the line $x = -\frac{mG_0}{a+m}$. We conclude that the attraction basin of the origin contain the regions (I), (II), (IV) and the greatest part of the region (III). This basin consists of the intersection between the region situated on the right side of the stable manifold of the point P_2 and the set

$$\{(x,y) \in \mathbb{R}^2 \mid x \geq -\frac{mG_0}{a+m}\} \cup \{(x,y) \in \mathbb{R}^2 \mid$$

$$\mid -G_0 < x < -\frac{mG_0}{a+m}, \; y > \frac{-bx(x+G_0)}{(a+m)x + mG_0}\}.$$

Theorem 3. *In the set*

$$\{(x,y) \in \mathbb{R}^2 \mid x \geq -\frac{mG_0}{a+m}\} \cup \{(x,y) \in \mathbb{R}^2 \mid -G_0 < x < -\frac{mG_0}{a+m}, \; y \geq 0\}$$

there are no closed orbits (periodic solutions), $\forall a, b, c, d, m > 0$. If

$$ac(mc - d^2) + (bd + mc)^2 \geq 0 \tag{10}$$

then nor in the set

$$\{(x,y) \in \mathbb{R}^2 \mid -G_0 < x < -\frac{mG_0}{a+m}, \; y < 0\}$$

there are no closed orbits.

Proof. ¿From Theorem 2 in the set of global stability

$$\{(x,y) \in \mathbb{R}^2 \mid x > -\frac{mG_0}{a+m}\}$$

cannot exist closed orbits and since the field lines cross the vertical asymptote only from the left to the right, we infer that there are no closed orbits which intersect this asymptote. So, the equilibrium point $(0,0)$ cannot be surrounded by closed orbits. Since any closed orbit must contain in the interior at least one equilibrium point and the other equilibrium point P_2 is a saddle, we infer that there are no closed orbits. Then the only closed orbits can be homoclinic orbits. Using the Bendixson's criterion we have,

$$div(F,G) = -\frac{ayG_0}{(x+G_0)^2} - b - d = 0 \Longleftrightarrow y = -(\frac{b+d}{aG_0})^2 \cdot (x+G_0)^2,$$

and since the parabola $y = -(\frac{b+d}{aG_0})^2 \cdot (x + G_0)^2$ lies in the region : $x < 0, y < 0$, we infer that a homoclinic orbit, if exists, must intersect this parabola in this region. Since, $ac(mc - d^2) + (bd + mc)^2 > 0$, we infer that

$$y(x_2) - y_2 = \frac{cG_0}{d} \cdot [\frac{ac(mc - d^2) + (bd + mc)^2}{(ac + bd + mc)^2}] > 0,$$

which means that the point P_2 is inside of the parabola. Then, eventually homoclinic orbits from P_2 must intersect the parabola and then such homoclinic orbit passes through the attraction basin of the origin, which is impossible. Then, there are no closed orbits, neither homoclinic orbits in the set

$$\{(x, y) \in \mathbb{R}^2 \mid -G_0 < x < -\frac{mG_0}{a + m},\ y \geq 0\},$$

nor in the set

$$\{(x, y) \in \mathbb{R}^2 \mid -G_0 < x < -\frac{mG_0}{a + m},\ y < 0\}.$$

□

Remark 5. *The inequality (10) generally holds, for the majority of persons. If*

$$ac(mc - d^2) + (bd + mc)^2 < 0$$

then $y(x_2) < y_2$ and the point P_2 is outside of the parabola. Then, can exist homoclinic orbits which do not pass through the attraction basin and intersect the parabola. The clinical interpretation is : For a healthy person, the blood-glucose level cannot be periodic during a day (between the hours 6-24).

5. Comparison with Other Models

In comparisson with other models we can see that here we consider the hiperglycaemiant hormones too, and in these models are considered only the glucose level and the insulin level. Of course, we come back to the bidimensional system considering the difference level between of the hiperglycaemiant hormones and insulin. In a future work we will consider a system in three dimensions including the hiperglycaemiant hormones level.

As we know, there are not models for the glucose homeostasy which presents two equilibrium points. Here, we get two equilibrium points, one asymptotic stable and another unstable saddle point which correspond to the hypoglycaemy critical level (below this critical level can appear the coma). Other recent models are the followings. For instance, in [3], the system is,

$$\begin{cases} I'(t) = r_I[\frac{G(t)}{K_G} + a(1 - \frac{G(t)}{K_G}) - \frac{p \cdot I(t-h_p) + (1-p)I(t-h)}{K_I}] \cdot I(t) \\ G'(t) = r_G[1 + b(1 - \frac{I(t)}{K_I}) - \frac{G(t)}{K_G}] \cdot G(t), \end{cases}$$

and presents a periodic solution which oscillates around the asymptotic stable homeostasic value. There are experimental data which confirm this oscillations and other which infirm these oscillations. Most of the experimental data confirm the convergence of the glucose levels to the asymptotic stable homeostasic value.

Also other models can be found in [14] and [15]. The model from [15] is a generalisation of the model from [14], and is the following

$$\begin{cases} G'(t) = -f(G(t)) - g(G(t), I(t)) + b_7 \\ I'(t) = -p(I(t)) + q(L(G_t)), \quad t > 0 \\ G(0) = G_b + b_0, I(0) = I_b + b_0 b_3 \text{ sssi } G(t) = G_b, \forall t \in [-b_5, 0]. \end{cases}$$

This system has one equlibrium point which is asymptotic stable.

Finally, we can specify that our system is dedicated to the prevention of the disease and correspond to a still healthy man. The remarks and the clinical interpretations serve to this prevention. All the other glucose blood regulation models in the literature start from the glucose tolerance test and is concerned to help in the control of the diabetic disease.

References

[1] E.Ackermann, L.C.Gatewood, J.W.Rosevear, G.D.Molnar, Model studies of blood-glucose regulation, *Bull. of Math. Biophys.*, **27**, special issue, 1965, 21-37.

[2] H.T.Banks, Modelling of the control system in glucose homeostasis, in: *Lecture notes in Biomathematics*, vol. **6** (Modelling and Control in the Biomedical Sciences), Springer Verlag, Berlin, 1975.

[3] I.Basov, D.Švitra, A possibility of taking into consideration of insulin "age structure" for modelling blood glucose dynamics, *Informatica*, 2000, 11, no.1, 87-96.

[4] I.Basov, D.Švitra, Glycemia monitoring, *Nonlinear Analysis: Modelling and control*, **2**, IMI, 1998, Vilnius.

[5] I.Basov, M.Meilũnas, D.Švitra, Glycemia monitoring: the problem of exogenous insulin input, *Mathematical Modelling and Analysis*, **4**, 1999, Vilnius.

[6] R.N.Bergman, Y.Z.Ider, C.R.Bowden, C.Cobelli, Quantitative estimation of insulin sensitivity, *Am. J. Physiol.*, **236**, (1979), 667-677.

[7] M.Berger, D.Rodbard, Toward the implementation of computers in diabetes mellitus, *Diabetes, Nutrition & Metabolism*, vol.**4**, 1, 1991.

[8] E.Bernard-Weil, D.Claude, Simulation of the glucose tolerance test by the model for the regulation of agonistic-antagonistic couples. Bipolar control, *C.R. Acad. Sci. Paris, Ser.I*, **305**, (1987), 303-306.

[9] A.Bica, A mathematical model for the study of glycaemic homeostasy, Studia Univ. Babes-Bolyai, *Mathematica*, vol. XLVIII, no. 2, 2003, 13-20.

[10] C.Cobelli, G.Toffolo, E.Ferrannini, A model of glucose kinetics and their control by insulin, compartmental and noncompartmental approaches, *Math. Biosci.*, **58**, (1982), 27-60.

[11] C.Cobelli, K.Thomaseth, The minimal model of glucose disappearance: Optimal input studies, *Math. Biosci.*, **83**, (1987), 127-155.

[12] C.Cobelli, K.Thomaseth, Optimal equidose inputs and rile of measurement error for estimating the parameters of a compartmental model of glucose kinetics from continuous and discrete-time optimal samples, *Math. Biosci.*, **89**, no.2, (1988), 135-147.

[13] El-Shal, M.Shendy,Digital modelling and robust control of glucose homeostasis system, *Int. J.Syst. Sci.*, **20**, no.4, (1989), 575-586.

[14] A.De Gaetano, O.Arino, Mathematical modelling of the intravenous glucose tolerance test, *J. Math. Biol.*, **40**, no.2, (200), 136-168.

[15] Jiaxu Li, Yang Kuang, Bingtuan Li, Analysis of glucose-insulin interaction models with time delay, *Discrete and Cont. Dyn. Systems-Series B*, vol. **1**, no. 1, (2001), 103-124.

[16] D.W.Jordan, P.Smith, *Nonlinear ordinary differential equations,* Claredon Press, Oxford, 1987.

[17] M.Meilūnas, On the blood glucose dynamics modelling, *Math. Model. Anal.*, **3**, (1998), 136-139.

[18] L.S.Maergojz, B.N.Varava, V.T.Manchuk, Mathematical modelling and forecasting of homeostasis of glucose and blood sugars, *Sov. J. Autom. Inf. Sci.*, **18**, no.2, (1985), 60-65.

[19] G.Pacini, R.N.Bergman, MINMOD, a computer program to calculate insulin sensitivity and pancreatic responsitivity from the frequently sampled intravenous glucose tolerance test, *Comput. Methods Programs Biomed.* **23**, (1986), 113-122.

[20] G.W.Swan, An optimal control model of diabetes mellitus, *Bull. Math. Biol.*, **44**, (1982), 793-808.

[21] C.Simon, G.Brandenberger, M.Follenius, Ultradian oscillations of plasma glucose, insulin, and C-peptide in man during continuous enteral nutrition, *J. Clin. Endocrinol. Metab.*, **64**, (1987), 307-314.

[22] J.Sturis, K.S.Polonsky, E.Mosekilde, E.Van Cauter, Computer-model for mechanism underlying ultradian oscillations of insulin and glucose, *Am. J. of Physiol.* **260**,(1991), 801-809.

[23] Ch. Thum, H. Laube, K.E. Schroder, S. Raptis, E.F. Pfeiffer, Das kontinuierliche Blutzuckertagesprofil in Korrelation zum Seruminsulinbei idealgewichtigen und normalgewichtigen Stoffwechselgesunden, *Dtsch. Med. Waschr.*, Bd.100, 1975.

In: Focus on Evolution Equations
Editor: Gaston M. N'Guerekata

ISBN: 978-1-60021-342-7

Chapter 5

INVARIANT MANIFOLDS OF FULLY NONLINEAR EVOLUTION EQUATIONS

Du Duc Thang and Nguyen Van Minh*
Department of Mathematics, Hanoi University of Science,
Khoa Toan, Dai Hoc Khoa Hoc Tu Nhien,
334 Nguyen Trai, Hanoi, Vietnam

Abstract

This paper is concerned with the existence of invariant manifolds of fully nonlinear evolution equations in Banach spaces of the form

$$\frac{du(t)}{dt} = Au(t), \quad t \geq 0,$$

where the operator A is assumed to be quasi m-accretive, proto-differentiable at the stationary point and generate a C_0-semigroup. The assumption on the proto-differentiability of A allows us to linearize the equations via their Yosida approximation. As a result we obtain conditions for the existence of invariant manifolds that extend the existsting ones to a general class of evolution equations with multi-valued operators. We conclude with some example in PFDEs and PDEs.

1. Introduction and Preliminaries

Let us consider the following fully nonlinear evolution equation in a Banach space $\mathbb{X}$

$$\frac{du(t)}{dt} = Au(t), \quad t \geq 0, \tag{NE}$$

where A is in general, a nonlinear multi-valued operator which generates a strongly continuous semigroup of nonlinear operators on a Banach space $\mathbb{X}$ and is Proto-differentiable.

The existence of invariant manifolds to differential equations has been known for a century. It is a central concern in the theory of differential equations because of its applications, as a powerful tool, in the study of nonlinear equations. The reader is referred to

*E-mail address: nguyenvm@jmu.edu

any books on nonlinear differential equations for the history of the problem (see, e.g., [12], [19] and the references therein). The results on the existence of invariant manifolds so far are mainly concerned with equations that can be linearized by using Frechet derivatives or so. Inspired by the approach of N. Kato [14] to stability linearization of fully nonlinear evolution equations, in our paper we make an attempt to study the existence of invariant manifolds of these equations via the concept of proto-derivative. This concept is general enough to linearize an operator that is multi-valued and non Frechet differentiable. The assumption on the proto-differentiability of A allows us to linearize the equations via their Yosida approximation. As a result, by using Hadamard's method of graphic transforms, we obtain conditions for the existence of invariant manifolds that extend the existsting ones to a general class of evolution equations with multi-valued operators.

We conclude this section by recalling some notions and notations. Let $\mathbb{X}$ and $\mathbb{Y}$ be Banach spaces and let $F : \mathbb{X} \to 2^{\mathbb{Y}}$ be a multi-valued operator. We define the domain, range and graph of F by

$$\begin{aligned}
\mathcal{D}(F) &= \{x \in \mathbb{X}, F(x) = \emptyset\}, \\
\mathcal{R}(F) &= \bigcup_{x \in \mathcal{D}(F)} F(x), \\
\mathcal{G}(F) &= \{(x, y) \in \mathbb{X} \times \mathbb{Y} : x \in \mathcal{D}(F),\ y \in \mathcal{R}(F)\},
\end{aligned}$$

For $(x, y) \in \mathcal{G}(F)$, let us define multi-valued operators $\partial_i F(x, y)$ and $\partial_s F(x, y)$ from $\mathbb{X}$ to $2^{\mathbb{Y}}$ by

$$\begin{aligned}
\mathcal{G}(\partial_i F(x, y)) &= \liminf_{t \downarrow 0} t^{-1}(\mathcal{G}(F) - (x, y)), \\
\mathcal{G}(\partial_s F(x, y)) &= \limsup_{t \downarrow 0} t^{-1}(\mathcal{G}(F) - (x, y)).
\end{aligned}$$

where lim inf and lim sup are taken in the sense of set convergence (cf.[22]). In other words,

$$\begin{aligned}
(u, v) \in \mathcal{G}(\partial_i F(x, y)) &\iff \forall t_n \downarrow 0, \exists (u_n, v_n) \to (u, v) \text{ in } \mathbb{X} \times \mathbb{Y} : \\
&\qquad (x + t_n u_n, y + t_n v_n) \in \mathcal{G}(F) \\
(u, v) \in \mathcal{G}(\partial_s F(x, y)) &\iff \exists t_n \downarrow 0, \exists (u_n, v_n) \to (u, v) \text{ in } \mathbb{X} \times \mathbb{Y} : \\
&\qquad (x + t_n u_n, y + t_n v_n) \in \mathcal{G}(F).
\end{aligned}$$

In the sense of graph inclusion, we have $\partial_i F(x, y) \subset \partial_s F(x, y)$. When $\partial_i F(x, y) = \partial_s F(x, y)$, we denote it by $\partial F(x, y)$ and say that F is *proto-differentiable* at x relative to y, $\partial F(x, y)$ is called *proto-derivative* of F at x relative to y. When F is single valued operator, we write $\partial F(x) = \partial F(x, F(x))$, and call proto-deriavative of F at x.

Definition 1.1. Let $(X, |\cdot|)$ be a Banach space. A (possibly multi-valued) operator $Q : X \to 2^X$ is said to be *accretive* if

$$|x - x^* + \lambda(y - y^*)| \geq |x - x^*|, \quad \forall (x, y), (x^*, y^*) \in \mathcal{G}(Q) \quad \text{and} \quad \lambda > 0.$$

An accretive oprator Q is called *m-accretive* if the range condition $\mathcal{R}(I + \lambda Q) = X$ is sastified for all (equivalently, for some) $\lambda > 0$, where I is the identity operator. If $Q + \omega I$

is m-accretive for some $\omega \in \mathbb{R}$ then we define the resolent and the Yosida approximation of Q by

$$J_\lambda^Q = (I + \lambda Q)^{-1}, \quad \text{and} \quad Q_\lambda = \frac{1}{\lambda}(I - J_\lambda^Q) \quad \text{for } \lambda > 0 \text{ with } \lambda\omega < 1.$$

Recall that $Q+\omega I$ is a m-accretive operator if and only if J_λ^Q is a single-valued operator with range $\mathbb{X}$, and satisfies

$$\|J_\lambda^Q x - J_\lambda^Q y\| \leq \frac{1}{1 - \lambda\omega}\|x - y\|$$

for all $\lambda\omega < 1; x, y \in \mathbb{X}$.

Definition 1.2. A linear semigroup $T(t)$ is said to have *exponential trichotomy* if there exist bounded projections $P_1,\ P_2,\ P_3$ on $\mathbb{X}$ and positive constant $N, \alpha, \beta, \alpha < \beta$, satisfying

(i) $P_1 + P_2 + P_3 = I$,

(ii) $P_j T(t) = T(t) P_j,\ j = 1, 2, 3,$

(iii) $T(t)_{ImP_j},\ j = 2, 3$, are homeomorphism from ImP_j onto $ImP_j,\ j = 2, 3, \forall t \geq 0$, (so $T(-t)P_j$ is defined as the inverse of $T(t)P_j$);

(iv) and the following estimates hold

$$\begin{aligned} \|T(t)P_1 x\| &\leq Ne^{-\beta t}\|P_1 x\|, && (\forall t \geq 0, x \in \mathbb{X}), \\ \|T(-t)P_2 x\| &\leq Ne^{-\beta t}\|P_2 x\|, && (\forall t \geq 0, x \in \mathbb{X}), \\ \|T(t)P_3 x\| &\leq Ne^{\alpha|t|}\|P_3 x\|, && (\forall t \in \mathbb{R}, x \in \mathbb{X}). \end{aligned}$$

In case the projector $P_3 = 0$, the semigroup $(T(t))_{t\geq 0}$ is said to have an exponential dichotomy.

2. Main Results

First, we state the following hypotheses on a stationary solution $\bar{u}$ of Eq. (NE):

H1 Forall $x \in \mathcal{D}(A)$, there exists a single-valued linear operator $\partial A(x) : \mathcal{D}(A) \subset X \to X$ such that $\partial A(x) + \omega I$ is m-accretive and

$$\mathcal{G}(\partial A(x)) = \lim_{t \downarrow 0} t^{-1}[\mathcal{G}(A) - (x, Ax)].$$

H2 There exists $\lambda_{\bar{u}} > 0$ and nondecreasing function $\epsilon_0 > 0$ such that

$$|J_\lambda^{\partial A(x)} v - J_\lambda^{\partial A(z)} v| \leq \lambda\epsilon_0 |v||x - z|$$

for $0 < \lambda < \lambda_{\bar{u}},\ x, z \in \mathcal{D}(A)$ and $v \in X$.

With the above hypotheses we consider the following equations

$$\begin{cases} \frac{d}{dt}u_\lambda(t) = A_\lambda u_\lambda(t), \\ u_\lambda(0) = x, \end{cases} \qquad x \in X, 0 \le t \le T, T \ge 0, \tag{E$_\lambda$}$$

where A_λ is the Yosida Approximation of A. We have that A_λ is a (nonlinear) Lipschitz continuous operator, which generates a C_0-semigroup $(S_\lambda(t))_{t\ge 0}$. Then $u_\lambda(t) = S_\lambda(t)x$, $x \in \mathbb{X}$ is the unique classical solution of Eq. (E $_\lambda$).

By (H1), A is proto differentiable and $\partial A(J_\lambda u) + \omega I$ is m-accretive. By [14, Lemma 1.2], we have

$$\begin{aligned}(\partial A(J_\lambda u))_\lambda &= \partial A_\lambda(J_\lambda u + \lambda A u) \\ &= \partial A_\lambda((J_\lambda + \lambda A)u) = \partial A_\lambda(u),\end{aligned}$$

The proto derivative operator $(\partial A(J_\lambda u))_\lambda$ is linear and Lipschitz continuous. Hence by [14, Lemma 1.1], the Yosida approximation A_λ of A is Gateaux differentiable and $dA_\lambda(u) = \partial A_\lambda(u)$, $\forall u \in \mathcal{D}(A)$. We put $\partial A_\lambda(t)x = \partial A(J_\lambda S_\lambda(t)x)$ for the simplicity of notation. For $0 < \lambda < \min\left\{\frac{1}{2\omega}, \lambda_{\bar{u}}\right\}$, $0 < \bar{r} < \left(\frac{r}{2}\right)^{-2\omega T}$, $x \in \mathbb{X}$, we have $|S_\lambda(t)x - \bar{u}| < r/2$ for all $0 \le t \le T$.

Now we consider the linearized equation for Eq. (E $_\lambda$)

$$\begin{cases} \frac{d}{dt}v_\lambda(t) = dA_\lambda(S_\lambda(t)x)v_\lambda(t), & 0 \le t \le T, \\ v_\lambda(0) = w, & w \in X. \end{cases} \tag{L$_\lambda$; w}$$

By (H1), $\partial A_\lambda(t)x + \omega I$ is m-accretive, then it satisfies the following estimate

$$\left\| J_\lambda^{\partial A_\lambda(t)x} v - J_\lambda^{\partial A_\lambda(t)x} w \right\| \le \frac{1}{1-\lambda\omega}\|v - w\|, \quad v, w \in X. \tag{1}$$

Also , by (H2), since $J_\lambda S_\lambda(t)x \in \mathcal{D}(A)$,

$$\left\| J_\lambda^{\partial A_\lambda(t)x} v - J_\lambda^{\partial A_\lambda(\tau)x} v \right\| \le \frac{\lambda}{1-\lambda\omega}\epsilon_0 \|v\| \|S_\lambda(t)x - S_\lambda(\tau)x\|. \tag{2}$$

Recall that ([19]) the semigroups $(S(t))_{t\ge 0}$ and $(T(t))_{t\ge 0}$ on a Banach space $\mathbb{X}$ are said to be ϵ-close (with exponent μ) if there are positive constants μ, η such that $\eta e^{\mu} < \epsilon$ and

$$\|\phi(t)x - \phi(t)y\| \le \eta e^{\mu(t)}\|x - y\|, \quad \forall t \ge 0 x, y \in \mathbb{X}, \tag{3}$$

where

$$\phi(t)x := S(t)x - T(t)x, \quad \forall t \ge 0, x \in \mathbb{X}.$$

Theorem 2.1. *Let $\bar{u}$ be the stationary solution of* (NE) *and let the assumptions (H1) and (H2) be made. Then the C_0-semigroups $(S(t))_{t\ge 0}$ and $(T(t))_{t\ge 0}$ generated by A and $\partial A(\bar{u})$, respectively, are ϵ_0-close with exponent μ_0 that depends only on the given operator A.*

Proof. Set

$$\phi(t)x = S(t)x - T(t)x, \tag{4}$$
$$\phi_\lambda(x) = \mathcal{S}_\lambda(x) - \mathcal{T}_\lambda(x), \tag{5}$$

where, for a given positive number T,

$$(\mathcal{S}_\lambda x)(t) = S_\lambda(t)x, \quad \forall t \in [0, T],$$
$$(\mathcal{T}_\lambda x)(t) = T_\lambda(t)x, \quad \forall t \in [0, T].$$

For easily calculations, let us take $\bar{u} = 0$. In order to prove that $(S(t))_{t\geq 0}$ and $(T(t))_{t\geq 0}$ are ϵ_0-close, it suffices to show that $Lip(\phi_\lambda) < \epsilon$, where $\epsilon = \epsilon(\epsilon_0)$ is positive and independent of x, y, i.e. the following estimate

$$\|\phi_\lambda(x) - \phi_\lambda(y)\| \leqslant \epsilon\|x - y\|, \quad \forall x, y \in \mathbb{X} \tag{6}$$

holds. Then, by letting $\lambda \to 0$, for fixed t, we have

$$\|\phi(t)x - \phi(t)y\| \leq \epsilon\|x - y\|, \quad \forall x, y \in \mathbb{X}; t \in [0, T]. \tag{7}$$

Our task is now to prove (6). By [14, lemma 3.4], we have

$$\begin{aligned}\|\phi_\lambda(x) - \phi_\lambda(y)\| &= \|\mathcal{S}_\lambda(x) - \mathcal{S}_\lambda(y) - \mathcal{T}_\lambda(x) + \mathcal{T}_\lambda(y)\| \\ &\leq \int_0^1 \|d\mathcal{S}_\lambda(\theta x + (1-\theta)y)(x - y) - d\mathcal{T}_\lambda(\theta x + (1-\theta)y)(x - y)\|d\theta. \end{aligned} \tag{8}$$

Let $\eta = \theta x + (1-\theta)y$. Then $v_\lambda^\eta(t) = [d\mathcal{S}_\lambda(\eta)(x-y)](t)$ and $v_\lambda^0(t) := v_\lambda(t) = [d\mathcal{T}_\lambda(\eta)(x-y)](t)$ are solutions of (E$_\lambda$) with the operators are $dA_\lambda(S_\lambda(t)\eta)$ and $dA_\lambda(0)$ respectively, and $w = x - y$. By lemma [14, lemma 3.5],

$$\begin{aligned}&\|d\mathcal{S}_\lambda(\theta x + (1-\theta)y)(x - y)(t) - d\mathcal{T}_\lambda(\theta x + (1-\theta)y)(x - y)(t)\| \\ &\qquad = \|v_\lambda^\eta(t) - v_\lambda(t)\| \leq \frac{1}{1-\lambda\omega} e^{\mu t} t\|\eta\|\epsilon_0 e^{\mu T}\|x - y\|. \end{aligned} \tag{9}$$

Then

$$\begin{aligned}\|\phi_\lambda(x) - \phi_\lambda(y)\| &\leq \int_0^1 \frac{e^{\mu t} t}{1-\lambda\omega}\epsilon_0 e^{\mu T}\|x - y\|\|\eta\|d\theta \\ &\leq \frac{e^{\mu T} T}{1-\lambda\omega} e^{\mu T}\epsilon_0\|x - y\| \int_0^1 \|\eta\|d\theta \\ &= \frac{T}{1-\lambda\omega} e^{2\mu T}\epsilon_0\|x - y\| \int_0^1 \|\theta x + (1-\theta)y\|d\theta \\ &\leq \frac{MTe^{2\mu T}}{1-\mu\lambda}\epsilon_0\|x - y\| := \epsilon(\epsilon_0)\|x - y\|, \end{aligned}$$

for $\left|\int_0^1 \|\theta x + (1-\theta)y\|d\theta\right| \leq M < +\infty$, and $\epsilon(\epsilon_0) = (1-\lambda\omega)^{-1}e^{2\mu T}MT\epsilon_0$. Therefore, we have (6). This completes the proof. □

2.1. The Case of Exponetial Dichotomy

Let us consider equation (NE).

Theorem 2.2. *Assume that the hypothese (H1) and (H2) are sastified, and $u = 0$ is the stationary solution of Eq.* (NE). *Moreover, assume that the strongly continuous semigroup $(T(t))_{t\geq 0}$ has an exponential dichotomy. Then there exists a positive constant δ_0, such that if $0 < \epsilon_0 < \delta_0$ Eq. (NE) has an unstable invariant manifold $W^u \subset \mathbb{X}$, presented as graph of a Lipschitz continuous mapping $g : \mathbb{X}^2 \to \mathbb{X}^1$ with small Lipschitz constant, i.e. $S(t)W^u = W^u$, $t \geq 0$, and $\lim_{\delta_0 \to 0} Lip(g) = 0$.*

Proof. The theorem is an immediate consequence of [19, Lemmas 2.11, 2.12, 2.13] and Theorem 2.1. □

Theorem 2.3. *With the assumptions in Theorem 2.2, the set*

$$W^s = \left\{ x \in \mathbb{X} : \lim_{t \to +\infty} S(t)x = 0 \right\} \tag{10}$$

is a stable invariant manifold of Eq. (NE), *represented by the graph of a Lipschitz continuous mapping $g : \mathbb{X}^1 \to \mathbb{X}^2$, i.e. $W^s = \mathcal{G}(g)$ and $S(t)W^s \subset W^s$, forall $t \geq 0$.*

Proof. The theorem is an immediate consequence of [19, Theorem 2.16] and Theorem 2.1 □

2.2. The Case of Exponetial Trichotomy

Theorem 2.4. *Assume that the hypothese (H1) and (H2) hold. Moreover, assume that the C_0-semigroup $(T(t))_{t\geq 0}$ generated by proto derivative $\partial A(\bar{u})$ has an exponetial trichotomy with positive constants K, α, β, $\alpha < \beta$ and families of projections $\{P_j,\ j = 1, 2, 3\}$. Then, there is a positive constant δ_0 such that for $0 < \epsilon_0 < \delta_0$, there exist invarinant manifolds, called center, center-unstable and stable invariant manifolds for equation* (NE), *represented respectively by Lipschitz continuous mappings*

$$k : ImP_3 \to ImP_1 \oplus ImP_2,\ g : Im(P_2 + P_3) \to ImP_1,\ h : ImP_3 \to ImP_2,$$

and $\lim_{\delta_0 \to 0} \max(Lip(k), Lip(g), Lip(h)) = 0$.

Proof. The theorem is an immediate consequence of [19, Theorems 2.19, 2.23, 2.24] and Theorem 2.1. □

3. Applications

Example 3.1. Consider the semilinear equation

$$\frac{d}{dt}u(t) = (L + F)u(t), \tag{11}$$

where $L : \mathbb{X} \to \mathbb{X}$ is the infinitesimal generator of a C_0 semigroup $(T(t))_{t\geq 0}$ satisfying $\|T(t)\| \leq Me^{\omega t}$, $F : \mathbb{X} \to \mathbb{X}$ is a nonlinear Fréchet differentiable operator, and $\bar{u} = 0$

is the stationary solution of Eq. (11). Then if the C_0 semigroup generated by the linear operator $L+DF(0)$ has an exponential dichotomy, there exists stable and unstable invariant manifolds for Eq. (11). Let $r_0 > 0$ fixed, consider the cut-off function, defined as

$$F_0(x) = \begin{cases} F(x) & \text{if} \quad \|x\| \leq r_0, \\ F\left(r_0 x/\|x\|\right) & \text{if} \quad \|x\| > r_0. \end{cases}$$

One can check that $F_0(x)$ is a Fréchet continuously differentiable in $B_{r_0}(0)$ with Lipschitz constant $Lip(F_0)$. Set $Au = (L + F_0)u$, $u \in \mathcal{D}(A) = \mathcal{D}(L)$. The operator L can beassumed to be m-accretive and generates a C_0 semigroup operator of contraction in the sense of Crandall-Liggett $(T(t))_{t\geq 0}$. From the proof of [14, corollary 2.2], A satisfies hypothesis (H1) and (H2), and then generates a semigroup $(S(t))_{t\geq 0}$. Our aim is to prove that two above semigroups are ϵ - close. Set

$$\phi(t)x = S(t)x - T(t)x.$$

$u(t) = S(t)x$ is the solution of Eq. (11), with F replaced by F_0, so it satisfies

$$S(t)x = T(t)x + \int_0^t T(t-s)F_0 S(s)x ds.$$

We then have the folowing estimates

$$\begin{aligned}
\|\phi(t)x - \phi(t)y\| &= \left\| \int_0^t (T(t-s)F_0 S(s)x - T(t-s)F_0 S(s)y)ds \right\| \\
&\leq \int_0^t \|(T(t-s)\| \|F_0 S(s)x - F_0 S(s)y)\| ds \\
&\leq Me^{\omega(t-s)} \int_0^t 2c(r)\|S(s)x - S(s)y\| ds \\
&\leq 2Mc(r)e^{\omega t} \int_0^t \|x-y\| ds \\
&\qquad\qquad (\text{for } \|S(s)x - S(s)y\| \leq e^{\omega s}\|x-y\|) \\
&\leq 2MTc(r)e^{\omega T}\|x-y\|,
\end{aligned}$$

for all $0 \leq t \leq T$, $x, y \in B_r(0)$, $0 < r \leq r_0$. As is well known, if the radian r is small enough, $c(r)$ will be so, and $\epsilon_1 = 2MTc(T)e^{\omega T}$ is also small enough. Our conclusion is that the Lipschitz constant of $\phi(1)$ is small, i.e. for all $0 < \epsilon < \epsilon_1$, $Lip(\phi(1)) \leq 2Mc(r)e^{\omega} < \epsilon$. Therefore, Eq. (11) satisfies theorem 2.1. According to the above theorems, the existence of stable and unstable invariant manifolds for the equation holds.

(2) Consider the KdV equation

$$\begin{cases} u_t + \epsilon u_{xxxx} + u_{xxx} - \eta u_{xx} + \gamma u + u u_x = f, \\ u(t,0) = u(t,2\pi), \end{cases} \tag{12}$$

where ϵ,η, γ are fixed positive numbers. Denote $H = L^2[0, 2\pi]$ and set $H^k_{2\pi} \subset H$ is the set of all 2π-periodic functions which has generation derivative order k. Assume $f \in H^3_{2\pi}$, let

$A_\epsilon u = \epsilon u_{xxxx} + u_{xxx} - \eta u_{xx} + \gamma u$, $\mathcal{D}(A_\epsilon) = H^2_{2\pi}$, and $F(u) = f - uu_x$. Using [15, theorem 3.2.4], we have A_ϵ is strong elliptic operator of order four in $\mathcal{D}(A_\epsilon)$, so it is sectorial operator. Our equation is now

$$u_t + A_\epsilon u = F(u). \tag{13}$$

For A_ϵ is a linear sectorial operator, it generates an analytic C_0-semigroup (see, e.g. [21]). Assume that $F(u)$ is Lipschitz continuous on $H^2_{2\pi}$ with small Lipschitz constant. Then by [14, theorem 4.1-4.3,4.5], the operator $\mathcal{A} + \omega I = -(A_\epsilon + F) + \omega I$ is m-accretive with $\omega = Lip(F)$. Therefore, the equation (12) satisfies hypotheses (H1) and (H2). By theorem 2.1, we conclude that C_0 semigroups $(T(t))_{t\geq 0}$ and $(S(t))_{t\geq 0}$ generated by $-A_\epsilon$ and $\mathcal{A}$ are ϵ_0-closed, for all small ϵ_0. Then the existence of invariant manifolds of KdV equation followed by the above theorems.

Example 3.2 (Age-dependent population dymanics). Let us consider the equation

$$\begin{cases} \frac{\partial}{\partial t}u(t,a) + \frac{\partial}{\partial a}u(t,a) = G(u(t,\cdot))(a), & t \geq 0,\ a \geq 0, \\ u(t,0) = F(u(t,\cdot)), & t \geq 0, \end{cases} \tag{14}$$

Let $(E, |\cdot|)$ be a Banach space and $L^1 = L^1(0,+\infty;E)$ be a Bochner integrable function space. Given two mappings $F : L^1 \to E$ and $G : L^1 \to L^1$, with F and G are continuously Fréchet differentiable, i.e.,

(F) For any $\phi \in L^1$, there exists a $DF(\phi) \in \mathcal{L}(L^1, E)$ such that

$$F(\phi + h) = F(\phi) + DF(\phi)h + o_F(h), \quad h \in L^1$$

where $o_F : L^1 \to E$, $|o_F(h)| \leq b_F(r)\|h\|_{L^1}$ for $\|h\|_{L^1} \leq r$ and $b_F : [0,\infty) \to [0,\infty)$ is a continuous increasing function satisfying $b_F(0) = 0$, and there exists a continuous increasing function $d_F : [0,\infty) \to [0,\infty)$ such that

$$\|DF(\phi) - DF(\psi)\|_{\mathcal{L}(L^1,E)} \leq d_F(r)\|\phi - \psi\|_{L^1}, \quad \text{for} \quad \|\phi\|_{L^1}, \|\psi\|_{L^1} \leq r.$$

(G) For any $\phi \in L^1$, there exists a $DG(\phi) \in \mathcal{L}(L^1, L^1)$ such that

$$G(\phi + h) = G(\phi) + DG(\phi)h + o_G(h), \quad h \in L^1$$

where $o_G : L^1 \to L^1$, $\|o_G(h)\|_{L^1} \leq b_G(r)\|h\|_{L^1}$ for $\|h\|_{L^1} \leq r$ and $b_G : [0,\infty) \to [0,\infty)$ is a continuous increasing function satisfying $b_G(0) = 0$, and there exists a continuous increasing function $d_G : [0,\infty) \to [0,\infty)$ such that

$$\|DG(\phi) - DG(\psi)\|_{\mathcal{L}(L^1,L^1)} \leq d_G(r)\|\phi - \psi\|_{L^1}, \quad \text{for} \quad \|\phi\|_{L^1}, \|\psi\|_{L^1} \leq r.$$

Let $\bar{u}$ is the stationary solution of Eq. (14), i.e., $\bar{u} \in W^{1,1} = W^{1,1}(0,\infty;E)$, $\bar{u}(0) = F(\bar{u})$ and $\bar{u}' = G(\bar{u})$, where $'$ stand for d/da.

References

[1] B. Aulbach, N. V. Minh (1996), "Nonlinear semigroups and the existence and stability of solutions of semilinear nonautonomous evolution equations", *Abstract and Applied Analysis*, **1** 351-380.

[2] A. V. Babin, G. Sell (2000), "Attractors of non-autonomous parabolic equations and their symmetry properties",*J. Diff. Equ.* **160** , 1-50.

[3] V. Barbu (1976),*Nonlinerar semigroups and differential equations in Banach spaces* , Noordhoff, Groningen.

[4] P. Bate, K. Lu, C. Zeng (2000), "Invariant foliation near normally hyperbolic invariant manifolds for semiflows",*Trans. Amer. Math. Soc.* **352**, no 10, 4641-4676.

[5] P. Bate, K. Lu, C. Zeng (1998),Existence and persistence of invariant manifolds for semiflows in Banach spaces, *Mem. Amer. Math. Soc.* **135**, no 645.

[6] J. Carr (1981),Applications of center manifold theory, *Applied Mathematical Sciences*, vol. 35, Springer Verlag, New York.

[7] X.-Y. Chen, J. K. Hale, B. Tan (1997), "Invariant foliations for C^1 semigroups in Banach spaces" ,*J. Diff. Equ.* **139**, 283-318.

[8] C. Chicone, Y. Latushkin (1997), "Center manifolds for infinite dimensional nonautonomous differential equation",*J. Diff. Equ.* **141**, 356-399.

[9] S. Chow, K. Lu (1988), "Invariant manifolds for flows in Banach spaces", *J. Diff. Equ.* **74** , 355-385.

[10] M. G. Crandall, T. M. Liggett(1971), "Generation of semi-groups of nolinear transformations on general Banach spaces",*Amer. J. Math.*, **93** , 265-298.

[11] J. K. Hale (1988),*Asymptotic behavior of dissipative systems* , Amer. Math. Soc., Providence, RI.

[12] D. Henry(1981),Geometric in theory of semilinear parabolic equations, Lecture notes in Math., vol 840.

[13] M. Hirsch, C. Pugh, M. Shub (1977), Invariant manifolds, *Lecture notes in Math.*, vol. 583, Springer .

[14] N. Kato (1995), "A principle of linearized stability for nonlinear evolution equations", *Trans. Amer. Math. Soc.* **347**, 2851-2868.

[15] A. Lunardi (1995), Analytic semigrpoups and optimal regularity in parabolic problems, *Progress in Nonlinear Differential Equations and Their Applications,* vol. 16, Birkhauser, Basel.

[16] C. Keller (1983), "On the stable manifolds for the nonliear wave equation with dissipation", *J. Diff. Equ.* **50**, 330-347.

[17] A. Kelley (1967), "The stable, center-stable, center, centerunstable and unstable manifolds", *J. Diff. Equ.* **3**, 540-570.

[18] X. Lin, J. So, J. Wu(1992), "Center manifolds for partial differential equations with delays",*Proc. Royal Soc. Edinbourg* **122 A** , 237-254.

[19] Nguyen Van Minh, J. Wu, Invariant manifolds of partial functional differential equations, *Journal of Differential Equations* **198** (2004), 381-421.

[20] S. Murakami, N. V. Minh (2002), " Some invariant manifolds for abstract functional diffrential equations and linearized stabilities", *Vietnam J. Math.*, **30** SI , 437-458.

[21] A. Pazy (1983), Semigroups of linear operators and applications to partial differential equations, *Applied Math. Sci.,* vol. 44, Springer.

[22] R. T. Rockafellar (1989), "Proto-differentiability of set-valued mappings and its applications in optimization ", *Ann. Inst. Poincaré Analyse Non Linéaire* **6**, 449-482.

[23] R. Rosa, R. Temam (1996), "Inertial manifolds and normal hyperbolicity", *Acta Applicandae Mathematicae*, **45**, 1-50.

[24] R. Sacker, G. Sell(1994), "Dichotomies for linear evolutionary equations in Banach spaces", *J. Diff. Equ.*, **78** , 17-67.

[25] J. C. Wells (1976), "Invariant manifolds of nonliear operators", *Pacific J. Math.* **62** , 285- 293.

In: Focus on Evolution Equations
Editor: Gaston M. N'Guerekata
ISBN: 978-1-60021-342-7

Chapter 6

GEVREY REGULARITY FOR SOME P.D.E GOVERNING THE MOTION OF A MICROPOLAR FLUID IN THE THREE DIMENSIONAL CASE*

Amor Kessab[†,‡]
Faculté de Mathématiques,
Université des Sciences et de la Technologie Houari Boumediène,
BP n°32, El-Alia 16111, Alger, Algerie

Abstract

We consider in this article the equations governing the movement of micropolar fluids in the three dimensional case . We establish a Gevrey regularity result like in the Navier-Stokes equations.

1991 Mathematics Subject Classification: Primary 35Q30; Secondary 76D05, 76D07, 76N10

Keywords and phrases: Navier-Stokes equations, Gevrey regularity

1. Introduction

In this work, we establish a result of Gevrey regularity for (weak) solutions for the following equations describing the movement of an incompressible micropolar fluid in the three dimension periodical case :

$$\begin{cases} \dfrac{\partial v}{\partial t} - (\mu + \alpha)\,\Delta v + (v.\nabla)\,v + \nabla p - 2\alpha \operatorname{rot} w = f_1 \\ \dfrac{\partial w}{\partial t} - (\gamma + \beta)\,\Delta w - (\varepsilon + \gamma - \beta)\ \nabla\,.\operatorname{div} w + 4\alpha w p - 2\alpha \operatorname{rot} v = f_2 \\ div(v) = 0. \end{cases} \tag{1.1}$$

*Communicated by Nguyen Van Minh

†E-mail address: amorkes@yahoo.fr

‡The author is grateful for professor Djamel Teniou, Mustapha Moulaï and anonymous referees for their helpful comments and corrections.

with the conditions :

$$\begin{cases} v(x,\ 0) = g_1(x) \\ w(x,\ 0) = g_2(x) \\ x \in \mathbb{R}^3 \end{cases} \tag{1.2}$$

$\mu, \varepsilon, \alpha,\ \beta$ and γ are physical parameters such that : μ, α, β and γ are positive, $3\varepsilon + 2\gamma > 0$; f_1 and f_2 represent respectively the massic force and the massic moment by unity of volume. The equations (1.1) are a generalization of Navier -Stokes equations (see [2]) .We look for $v(.,\ t)$, $w(.,\ t)$ and $p(.,\ t)$ supposed to be Q-periodical for all t in $\mathbb{R}_+$, where Q is a cube$(0,\ l)^3$, $l > 0$. This last condition will be replaced by boundary conditions in the case of a regular open;
the periodicity does not give rise to problems related to the boundary and let the term ∇p vanish.

Our present work is inspired from the article of Chen ([1]).

2. Notations and Definitions

a) If $H^m(Q)$denotes the Sobolev space of order m, we denote by $\mathbb{H}^m$ the space $(H^m)^3$ .We respectively denote by $(.,\ .)$ and $(.(.,\ .))$ the scalar products in $L^2(Q)$ and $\left(L^2(Q)\right)^3$and by $|.|, ||.||$ the corresponding norms. We denote by :

$$\mathbb{L}^2(Q) = \left(L^2(Q)\right)^3$$

b) Taking $l = 2\pi$ (for simplification) and using the Fourier series, we show that : $\mathbb{L}^2(Q) = \{u\ /\ u = \sum u_j e^{ijx}\ ,\ u_j \in C^3,\ u_{-j} = \overline{u}_j$ and $\sum_{j\in\mathbb{Z}^3} |u_j|^2 = \frac{1}{(2\pi)^3}||u||^2 < \infty\}$.

For this, see [4]. We set in the following :

$$H = \left\{u \in \mathbb{L}^2(Q) : j.u_j = 0,\ \forall j \in \mathbb{Z}^3\ and\ u_0 = 0\right\}.$$

H is then a closed subspace of $\mathbb{L}^2(Q)$.

Remark 2.1. $J.u_j = 0 \Leftrightarrow \operatorname{div}(u) = 0$ and $u_0 = 0$ comes from the fact that we can always suppose $\int_Q u_0 = 0$,when we deal with Navier-Stokes equations, see [4].

Definition 2.2. Let A be the operator defined on $D(A)$ into H by :

$$D(A) = \left\{u \in H / \sum |j|^4 . |u_j|^2 < \infty\right\}$$

and

$$A(u) = \sum |j|^2 u_j e^{ijx}, \ \forall u \in D(A).$$

Definition 2.3.
a) A is nothing else than the non bounded operator $-\Delta$ on H.
b) $D(A)$ is a Hilbert space equipped with the norm :

$$\|u\|_{D(A)} = ||Au||$$

and $A : D(A) \longrightarrow H$ is an isomorphism.(see for example [4])

Definition 2.4. Let $\alpha > 0$, we define the operator $A^\alpha : D(A) \longrightarrow H$ by :

$$D(A^\alpha) = \left\{ u \in H \ : \sum_{j \in \mathbb{Z}^3} |j|^{4\alpha} . |u_j|^2 = \frac{1}{(2\pi)^3} ||A^\alpha u||^2 < \infty \right\}$$

$$A^\alpha(u) = \sum |j|^{2\alpha} u_j e^{ijx}, \ \forall u \in D(A)$$

Remark 2.5. $(e^{ijx})_{j \in \mathbb{Z}^3}$ is an orthonormal basis of $L^2(Q)$ and $\left\{|j|^2\right\}_{j \in \mathbb{Z}^3 *}$ are the eigenvalues corresponding to the operator A.

Owing to the introduction of Fourier series as in [4], we have the characterization of the Sobolev space:

$$\mathbb{H}^m(Q) : \mathbb{H}^m(Q) = \left\{ u \ / \ u = \sum_{k \in \mathbb{Z}^3} u_k e^{ikx}, \ \overline{u}_k = u_{-k}, \ \sum_{k \in \mathbb{Z}^3} |k|^{2m} |u_k|^2 < \infty \right\}$$

Thanks to this characterization, it is clear that $||A^\alpha u||$ is equivalent to $\|u\|_{\mathbb{H}^{2\alpha}(Q)}$, which quantity will be later denoted by $||u||_\alpha$.

Notice that $||u||_m$ is equivalent to the norm: $\left\{ \sum_{k \in \mathbb{Z}^m} \left(1 + |k|^{2m}\right) . |u_k|^2 \right\}^{\frac{1}{2}}$.

3. The Gevrey class $\mathbf{D}\left(e^{\sigma A^\alpha}\right)$

Let α and σ be positive. We define :

$$D\left(e^{\sigma A^\alpha}\right) = \left\{ u \in H \ / \ \frac{1}{(2\pi)^3} \sum_{j \in \mathbb{Z}^3} e^{2\sigma |j|^{2\alpha}} . |u_j|^2 = \left|\left|e^{\sigma A^\alpha} u\right|\right|^2 < \infty \right\}$$

where $||.||$ has the same signification as in 2 and $e^{\sigma A^{\alpha}}u = \sum_{j \in Z^3} e^{\sigma |j|^{2\alpha}}.u_j e^{ijx}$.

Note that if $u \in D\left(e^{\sigma A^{\alpha}}\right)$, then $|u_j| < c.e^{-\sigma |j|^{2\alpha}}$ is exponentially decreasing.

We denote in the following :

$$\left\|e^{\sigma A^{\frac{\alpha}{2}}} u\right\| = |u|_{\sigma,\ \alpha} \text{ and } \left(e^{\sigma A^{\frac{\alpha}{2}}} u,\ e^{\sigma A^{\frac{\alpha}{2}}} v\right) = (u,\ v)_{\sigma,\ \alpha}$$

with these notations let's recall the result established in [1] for the equation :

$$\left\{\begin{array}{lll} \dfrac{\partial u}{\partial t} - \nu \Delta u + (u \cdot \bigtriangledown) u + \bigtriangledown p & = & f \\ \operatorname{div}(u) & = & 0,\ (x,\ t) \in \mathbb{R}^3 \times \mathbb{R} \\ u(x,\ 0) & = & u_0(x),\ x \in \mathbb{R}^3 \end{array}\right. \tag{3.3}$$

where $u(.,\ t)$and $p(.,\ t)$are $Q-$periodic for all t in $\mathbb{R}_+$.

(3.3) may be reduced to :

$$\left\{\begin{array}{lll} \dfrac{du}{dt} + \nu A u + B u & = & f \\ u(0) & = & u_0 \end{array}\right. \tag{3.4}$$

where B is a bilinear form defined by : $B(u,\ v) = P(u \cdot \bigtriangledown v)$, for all $u,\ v$ in $D(A)$, P being the orthogonal projection on H and $Bu = B(u, u)$. (see [4]).

Remark 3.1. Thanks to the fact that $\operatorname{div} u = 0$, the term $(u.\nabla) u$ writes as : $(u.\nabla) u = \sum_{i=1}^{3} \operatorname{div}(u_i.u) = \nabla.(u \bigotimes u)$. As $u \bigotimes u$ belongs to $\mathbb{L}^1$ whenever u is in $\mathbb{L}^2(Q)$, $(u.\nabla) u$ has a sense in $D'\left(\mathbb{R}^3\right)$.

We can now recall the theorems obtained by C. Foias-Guillope and R. Temam :

Theorem 3.2 ([3]). *Suppose* $u_0 \in H,\ f \in L^{\infty}\left(\mathbb{R}_+,\ D\left(A^{(m-1)/2}\right)\right)$ *for* $m \in \mathbb{N}^*$. *Then there exists at least a (weak) solution* u *of the problem (3.4) and an open* Ω *in* $\mathbb{R}_+$ *whose complementary is of measure zero such that :*

$u : \Omega \longrightarrow D\left(A^{m/2}\right)$ *is continuous.*

Further :

$$\int_0^T \frac{\left|A^{(m+1)/2} u(t)\right|^2}{\left(1 + \left|A^{m/2} u(t)\right|^2\right)^{\frac{2m}{2m-1}}} dt \leq K_1(1 + T)$$

where 'weak solution' is in the "Leray" sense and we recall this definition :

Definition 3.3 (see [1], [3] and [7]). Set $V = D\left(A^{1/2}\right)$ and V' the dual of V. Let $f \in L^2(0,T;V')$ and $u_0 \in H$.

A weak solution of (3.4) is a function u verifying : $u \in L^2(0,T;V)$ such that : $\dfrac{du}{dt} \in L^1(0,T;V')$ and :

$$\begin{cases} \dfrac{du}{dt} + \nu Au + Bu &= \quad f \text{ in } V' \\ u(0) &= \quad u_0 \end{cases}$$

Theorem 3.4 ([3], [5]). *Suppose* $u_0 \in D\left(A^{1/2}\right)$ *and* $f \in L^\infty\left(\mathbb{R}_+,\ D\left[e^{\delta A^{\frac{\alpha}{2}}}\right]\right)$ *for* $\delta > 0$ *and* $\alpha \in \left]0,\ 1\right[$. *Then there exists* T_1 *depending on* $\nu, Q, \delta, \|f\|_{L\infty}$ *and on* $|A|^{1/2}\, u_0$ *such that the problem (3.4) has an unique solution on* $[0,\ T_1]$, *verifying :*

$$[0,\ T_1] \longrightarrow H$$

$$t \longmapsto e^{\delta t A^{\frac{\alpha}{2}}}.A^{\frac{1}{2}} u(t) \quad \textit{continuous}$$

Theorem 3.5 (W. Chen). *Suppose that* $u_0 \in H,\ f \in L^\infty\left(\mathbb{R}_+,\ D\left[e^{\delta A^{\frac{\alpha_0}{2}}}\right]\right)$ *for given* $\delta > 0,\ \alpha_0 \in \left]0,\ 1\right[$. *Then there exists a (weak) solution* u *of (3.4), an open set* Ω *in* $\left]0,\ +\infty\right[$ *whose complementary is of measure zero, such that :*

a) $\forall \sigma > 0,\ 0 < \alpha < \alpha_0 : t \longrightarrow u(t)$ *is continuous from* Ω *into* $D\left(e^{\sigma A^{\frac{\alpha}{2}}}.A^{\frac{1}{2}}\right)$.

b) if $\alpha_0 > \alpha_* = \dfrac{\ln(5/3)}{\ln(2)}$, *we have :*

$$\int_0^T \frac{\left|e^{\sigma A^{\frac{\alpha_*}{2}}}.Au(t)\right|^2}{\left(1+\left|e^{\sigma A^{\frac{\alpha_*}{2}}}.A^{\frac{1}{2}}u(t)\right|^2\right)} dt \leq K\,(1+T)$$

where $K = K\left(v,\ Q,\ \|f\|,\ |u_0|\right)$ *and* $\|f\|_{L^\infty\left(\mathbb{R}_+,\, D\left(e^{\sigma A^{\frac{\alpha_*}{2}}}\right)\right)} = \|f\|$.

We write the problem (1.1) under the following form :

We set $G = (g_1,\ g_2)\,,\ U = (v, w),\ F = (f_1,\ f_2)$ and

$$\widetilde{\Delta} = \begin{bmatrix} \{-(\mu+\alpha)\,\Delta v + (v.\nabla)\,v - 2\alpha \operatorname{rot} w\}\,, \\ \qquad \{-(\gamma+\beta)\,\Delta w - (\varepsilon+\gamma-\beta)\,\nabla.\operatorname{div} w + 4\alpha w - 2\alpha \operatorname{rot} w\} \end{bmatrix}$$

$$\begin{cases} div^*(U) = div^*(v,\ w) &= \ div(v) \\ \nabla^* p &= \ (\nabla p,\ 0) \end{cases}$$

We take in the following $\gamma + \beta = \mu + \alpha = \nu.$, $\varepsilon + \gamma - \beta = 1$ and $\alpha = 1$.we finally set $\Delta^* = \frac{1}{\nu}\widetilde{\Delta}$. The problem (1.1) becomes then :

$$\begin{cases} \dfrac{\partial U}{\partial t} - \nu\Delta^* U - \nabla^* p = F \ \ sur\ \mathbb{R}^6 \times \mathbb{R}_+. \\ \operatorname{div}^* (U) = 0 \\ U(0) = G \ \ sur\ \mathbb{R}^6 \end{cases} \tag{3.5}$$

We also suppose that :

$$\int_Q g_1 = \int_Q g_2 = 0$$

which is no restriction, see for this [4]. Multiplying in (3.5) by $U' = \left(v', w'\right)$ in an appropriate space, and integrating by parts over a period,we show that: (see by example [4]) :

$$\frac{d}{dt}(U,\,U') + \left\{\int \nabla v \nabla v' + \int \nabla w \nabla w'\right\} + \{(S(U,\,U),\,U')\} = (F,\,U')$$

here:

$$(U,\,U') = \int U.U' \tag{3.6}$$

and

$$\{S(U,U),U'\} = \int (v.\nabla v).v' - 2\int \operatorname{rot} w.v' - \int \nabla.\operatorname{div} w.w' + 4\int w.w' - 2\int \operatorname{rot} w.w'$$

Unfortunately here $S(U,\,V)$ is not a bilinear form as in [1]. The problem consists next to give an estimation of this quantity in an appropriate Gevrey class. In (3.6) the term $\int (v.\nabla v).v'$ comes from the trilinear form :

$$b(u,\,v,\,w) = \int (u.\nabla v)w$$

and we can then , for the later estimations consider the bilinear form B already introduced:

$$\langle B(u,v),w\rangle = b(u,\,v,\,w)$$

because then $\int (v.\nabla v)\,v' = \langle B(v,\,v),\,v'\rangle$ and we use the estimations as in [1].

Set $B^*(U) = B^*(U,\ U)$ where : $\langle B^*(U), U'\rangle = \int (v.\nabla v)\, v'$ and

$\langle B^*(U,\ Y), U'\rangle = \int (v.\nabla y_1)\, v'$ where $Y = (y_1, y_2) \in \mathbb{R}^6$.

The problem (3.5) becomes then :

$$\begin{cases} \dfrac{dU}{dt} + \nu \mathcal{A}U + B^*(U) + S^\circ(U) = F \\ U(0) = G \\ \mathcal{A} = -(\Delta v,\ \Delta w) \end{cases} \tag{3.7}$$

where $S^\circ(U) = S(U,\ U)$, with :

$$\left(S^\circ(U),\ U'\right) = 4\int w.w' - 2\int \left[\text{rot}\, w.w' + \text{rot}\, w.v'\right] - \int \nabla.\, \text{div}(w).w' \tag{3.8}$$

4. Estimation of the Term $S^0(U)$ in the Gevrey Spaces

If $||.||$ is the norm on $\mathbb{L}^2(Q) = \left(L^2(Q)\right)^3$, we denote by $\|\ .\ |$ the norm on $\left(L^2(Q)\right)^6$ and let H as already introduced.

Definition 4.1. If $\mathcal{A}$ is the operator given in (3.7), we define the domain of $\mathcal{A}$ by :

$$D(\mathcal{A}) = \left\{(v, w) \in H \times \mathbb{L}^2(Q)\ \ /\sum |k|^4 \left(|v_k|^2 + |w_k|^2\right) = \|\mathcal{A}U|^2 < \infty\right\}$$

and

$$\mathcal{A}(U) = \sum_{k \in \mathbb{Z}^3} |k|^2 (v_k,\ w_k)\, e^{ikx},\ \forall U = (u,\ w) \in D(\mathcal{A}).$$

the term w_0 being supposed to be zero as in the remark 2.1.

Remark 4.2. $D(\mathcal{A})$ is a Hilbert space equipped with the norm :

$$\|U\|_{D(\mathcal{A})} = \|\mathcal{A}U|.$$

Definition 4.3. The operator $\mathcal{A}^\alpha,\ \alpha > 0$:
We define $\mathcal{A}^\alpha : D(\mathcal{A}^\alpha) \longrightarrow H \times \mathbb{L}^2(Q)$ by :

$$D(\mathcal{A}^\alpha) = \left\{ U = (v, w) \in H \times \mathbb{L}^2(Q)\ /\ \sum_{j \in \mathbb{Z}^3} |j|^{4\alpha} \left(|v_j|^2 + |w_j|^2\right) = \frac{1}{(2\pi)^3} \|\mathcal{A}^\alpha U|^2 < \infty \right\}$$

Remark 4.4. (for the notations see the previous pages and remark 2.5.
The characterization of $\mathbb{H}^m(Q)$ also gives :

a) $\|\mathcal{A}U| \simeq \|U\|_{\mathbb{H}^{2m}(Q)\times\mathbb{H}^{2m}(Q)}$

b) $\|U\|_m^2 = \sum_{k\in\mathbb{Z}^3}\left(1+|k|^{2m}\right)\left(|v_k|^2+|w_k|^2\right).$

Definition 4.5 (The Gevrey space $D\left(e^{\sigma\mathcal{A}^\alpha}\right)$). Let $\alpha,\ \sigma$ be positive, we define :

$$D\left(e^{\sigma\mathcal{A}^\alpha}\right)=\left\{U\in H\times\mathbb{L}^2(Q)\ /\ \sum e^{2\sigma|k|^{2\alpha}}\left(|v_k|^2+|w_k|^2\right) = (2\pi)^3\left\|e^{\sigma\mathcal{A}^\alpha}U\right|^2<\infty\right\} \tag{4.9}$$

where $\|.|$ has the same meaning as in the previous pages. Indeed :

$$\mathcal{A}^\alpha U=\sum|k|^{2\alpha}(v_k,\ w_k)\,e^{ikx}. \tag{4.10}$$

Set :

$$|U|_{\sigma,\ \alpha}=\|\ e^{\sigma\mathcal{A}^{\frac{\alpha}{2}}}U\ |$$

and $(U,\ V)_{\sigma,\ \alpha}$ the scalar product in the space $D\left(e^{\sigma\mathcal{A}^{\frac{\alpha}{2}}}\right)$:

$$(U,\ V)_{\sigma,\ \alpha}=\left(e^{\sigma\mathcal{A}^{\frac{\alpha}{2}}}U,\ e^{\sigma\mathcal{A}^{\frac{\alpha}{2}}}V\right)$$

Let $U,\ V,\ W$ be in $D\left(e^{\sigma\mathcal{A}^{\frac{\alpha}{2}}}\mathcal{A}\right)$, $\alpha\in\,]0,\ 1[$. Now the matter is to estimate, for $\mathcal{A}U$ in $D\left(e^{\sigma\mathcal{A}^{\frac{\alpha}{2}}}\right)$ that is for $U\in D\left(e^{\sigma\mathcal{A}^{\frac{\alpha}{2}}}\mathcal{A}\right)$, the quantity : $(S^\circ(U),\ \mathcal{A}U)_{\sigma,\ \alpha}$ where S° is as in (3.8). Remind first that if B is the aforementioned bilinear form , we have the following result in our case whose proof follows step by step that of [1].

Proposition 4.6. *Let $U,\ V,\ W\in D\left(e^{\sigma\mathcal{A}^{\frac{\alpha}{2}}}\mathcal{A}\right)$, $\alpha\in\,]0,\ 1[$, then :*

$$\left|(B^*(U,\ V),\ \mathcal{A}W)_{\sigma,\ \alpha}\right| \le c_1\left|\mathcal{A}^{\frac{1}{2}}U\right|^{\frac{1}{2}}_{\varepsilon_\alpha\sigma,\ \alpha}\cdot\left|\mathcal{A}^{\frac{1}{2}}V\right|_{\sigma,\ \alpha}\cdot|\mathcal{A}U|^{\frac{1}{2}}_{\varepsilon_\alpha\sigma,\ \alpha}\cdot|\mathcal{A}W|_{\sigma,\ \alpha}$$
$$+c_2\,|\mathcal{A}V|^{\frac{1}{2}}_{\varepsilon_\alpha\sigma,\ \alpha}\cdot\left|\mathcal{A}^{\frac{1}{2}}V\right|^{\frac{1}{2}}_{\varepsilon_\alpha\sigma,\ \alpha}\cdot\left|\mathcal{A}^{\frac{1}{2}}U\right|_{\sigma,\ \alpha}\cdot|\mathcal{A}W|_{\sigma,\ \alpha}$$

with: $\varepsilon_\alpha=2^\alpha-1$, $\varepsilon_\alpha\in\,]0,1[$. In the second term of the sum, we exchange U into V.

Remark now that :

a/ the term $\int ww'$ writes as :

$$\int ww' = \sum_{l,\, k \in \mathbb{Z}^3} w_k.w'_l.\int_Q e^{i(k+l)x}dx = (2\pi)^3 \sum w_k.w'_{-k}, \quad \text{(periodicity)}.$$

if $w = \sum_{k\in\mathbb{Z}^3} w_k e^{ikx},\ w' = \sum_{l\in\mathbb{Z}^3} w'_l e^{ilx},\ k,\ l \in \mathbb{Z}^3.$

b/ the term $\int \nabla.\operatorname{div}(w).w'$ is computed as :

First $\operatorname{div} w = i \sum_{k\in\mathbb{Z}^3} \langle k,\ w_k\rangle e^{ikx}$, then :

$$\int \nabla \operatorname{div}(w).w' = \int \sum_{k,\, l} \langle k,\ w_k\rangle . \left\langle w'_l,\ k\right\rangle e^{i(k+l)x}$$

and then :

$$\int \nabla \operatorname{div}(w).w' = (2\pi)^3 \sum \langle k,\ w_k\rangle . \left\langle k, w'_{-k} \right\rangle$$

c/ the term $\int \operatorname{rot} w.w'$:

By correctly writing $\operatorname{rot} w$ where :

$$w = \sum_{k\in\mathbb{Z}^3} \left(w_k^1,\ w_k^2,\ w_k^3\right) e^{ikx},\ w_k = \left(w_k^1,\ w_k^2,\ w_k^3\right)$$

we get :

$$\int \operatorname{rot} w.w' = i(2\pi)^3 \sum_{k\in\mathbb{Z}^3} \langle k,\ U_k\rangle,\ k \in \mathbb{Z}^3$$

where U_k is a vector whose components behave as the scalar product $w_k.w'_k$. More precisely :

$$U_k = \left(w_k^2.w'^3_{-k} - w_k^3.w'^2_{-k},\ w_k^3.w'^1_{-k} - w_k^1.w'^3_{-k},\ w_k^1.w'^2_{-k} - w_k^2.w'^1_{-k}\right)$$

For the term $\int \operatorname{rot} w.v'$we have a similar writing. The remarks (a), (b), and (c) allow now to use the proposition 4.6 to show exactly the same theorem of W. Chen already stated, with the non linear terms behaving as scalar products for the problem (3.7).

Remind that :

$$\begin{aligned}(S^{\circ}(U), \mathcal{A}U)_{\sigma,\,\alpha} &= \left(e^{\sigma\mathcal{A}^{\frac{\alpha}{2}}} S^{\circ}(U), \, e^{\sigma\mathcal{A}^{\frac{\alpha}{2}}} \mathcal{A}U\right) \\ &= \left(S^{\circ}(U), \, e^{2\sigma\mathcal{A}^{\frac{\alpha}{2}}} \mathcal{A}U\right)\end{aligned}$$

$\mathcal{A}U$ writes :

$$\mathcal{A}U = \sum_{j\in\mathbb{Z}^3} |j|^2 (v_j, \, w_j)\, e^{ijx}$$

and then

$$e^{2\sigma\mathcal{A}^{\frac{\alpha}{2}}}.\mathcal{A}U = \sum_{j\in\mathbb{Z}^3} e^{2\sigma|j|^{\alpha}} |j|^2 (v_j, \, w_j)\, e^{ijx} = (v', \, w')$$

the previous remark gives for the points a and c :

$$\int w.w' = (2\pi)^3 \sum_{k\in\mathbb{Z}^3} e^{2\sigma|k|^{\alpha}}. |k|^2 .w_k.w'_{-k} \tag{4.11}$$

and finally following the remark concerning (c) :

$$\int \operatorname{rot} w.w' = (2\pi)^3 i. \sum_{k\in\mathbb{Z}^3} \langle R_k, w'_{-k}\rangle . \tag{4.12}$$

where $R_k = \left(k_2 w_k^3 - k_3 w_k^2, \, k_3 w_k^1 - k_1 w_k^3, \, k_1 w_k^2 - k_2 w_k^1\right)$, with $w_k = \left(w_k^1, \, w_k^2, \, w_k^3\right)$ and $k = (k_1, \, k_2, \, k_3) \in \mathbb{Z}^3$. Consequently :

$$\int \operatorname{rot} w.w' = (2\pi)^3 . \sum_{k\in\mathbb{Z}^3} e^{2\sigma|k|^{\alpha}}. |k|^2 R_k.\overline{w'}_k$$

of course $\overline{w}_k = w_{-k}$. Finally :

$$\int \operatorname{rot} w.v' = (2\pi)^3 . \sum_{k\in\mathbb{Z}^3} e^{2\sigma|k|^{\alpha}}. |k|^2 .S_k.v_k. \tag{4.13}$$

Note that : $\|R_{\cdot k}\| \leq C\,|k|\,.\,|w_k|$ and $\|S_k\|$ verify a similar inequality. We have then :

$$\begin{aligned}\tfrac{1}{(2\pi)^3}\left(S^{\circ}(U), \, \mathcal{A}U\right)_{\sigma,\alpha} &= 4\sum_{k\in\mathbb{Z}^3} e^{2\sigma|k|^{\alpha}}. |k|^2 .w_k.w_{-k} - \sum_{k\in\mathbb{Z}^3} e^{2\sigma|k|^{\alpha}} \langle k, \, w_k\rangle . \langle k, w_{-k}\rangle \\ &\quad -2\sum_{k\in\mathbb{Z}^3} R_k e^{2\sigma|k|^{\alpha}}. |k|^2 .\overline{w}_k - 2\sum_{k\in\mathbb{Z}^3} e^{2\sigma|k|^{\alpha}}. |k|^2 .S_k.v_k\end{aligned}$$

Set $J_1 = 4\sum_{k\in\mathbb{Z}^3} e^{2\sigma|k|^{\alpha}}. |k|^2 .w_k.w_{-k} - \sum_{k\in\mathbb{Z}^3} e^{2\sigma|k|^{\alpha}} \langle k, \, w_k\rangle . \langle k, w_{-k}\rangle$

$$-2\sum_{k\in\mathbb{Z}^3} R_k e^{2\sigma|k|^\alpha}.\left|k\right|^2.\overline{w}_k$$

the first term in the previous sum and J_2 the second term .We then have the following proposition:

Proposition 4.7. *Let* $U = (v,w) \in D\left(e^{\sigma A^{\frac{\alpha}{2}}}.\mathcal{A}\right), \alpha \in \left]0,1\right[$. *The following estimates hold :*

a) $|J_1| \le c_0.\left|\mathcal{A}^{1l2}U\right|^2_{\sigma,\alpha}$.

b) $|J_2| \le c_\nu.\left|\mathcal{A}^{1l2}U\right|^2_{\sigma,\alpha} + \frac{\nu}{8}\left|AU\right|^2_{\sigma,\alpha}$.

Proof. a). Set: $\xi = \sum_{j\in\mathbb{Z}^3} |j|\,|w_j|\,e^{\sigma|j|^\alpha}.e^{ijx}$.

We have :

$$|J_1| \le c.\sum_{k\in\mathbb{Z}^3} |k|^2\,|w_k|^2\,e^{2\sigma|k|^\alpha} = \mu_\alpha$$

with the help of the following integral μ_α expresses as :

$$\mu_\alpha = \frac{1}{(2\pi)^3}\int_Q \xi.\xi dx.$$

which gives :

$\mu_\alpha \le c.\,|\xi|^2_{L^2(Q)}$.

Or

$$|\xi|_{L^2(Q)} = (2\pi)^3.\sum_{k\in\mathbb{Z}^3} |k|^2\,|w_k|^2\,e^{2\sigma|k|^\alpha} = \left|\mathcal{A}^{1l2}U\right|^2_{\sigma,\alpha}$$

that is to say (a) .

(b) taking in consideration (4.13) and the remark which follows it, we have :

$$|J_2| \le c_1.\sum_{k\in\mathbb{Z}^3}.\,|k|^3\,|w_k|\,.\,|v_k|\;\; e^{2\sigma|k|^\alpha}$$

Set : $\xi = \sum_{j\in\mathbb{Z}^3} |j|\,|w_j|\,e^{\sigma|j|^\alpha}.e^{ijx}$ the same as in (a) .

and $\eta = \sum_{j\in\mathbb{Z}^3} |j|^2\,|v_j|\,e^{\sigma|j|^\alpha}.e^{ijx}$.

We have then : $|J_2| \le c_2.\int_Q \xi.\eta dx. \le c_2.\,|\xi|_{L^2(Q)}\cdot|\eta|_{L^2(Q)}$ ·

But:

$|\xi|^2_{L^2(Q)} \le c_2.\left|\mathcal{A}^{1l2}U\right|^2_{\sigma,\alpha}$.

by the same way and by correctly computing $|\eta|_{L^2(Q)}$ we get :
$|\eta|_{L^2(Q)} \leq c_3. |AU|_{\sigma,\alpha}$.
Consequently: $|J_2| \leq c_4. |AU|_{\sigma,\alpha} \cdot |A^{1l2}U|_{\sigma,\alpha}$ which obviously gives (b).

□

we now have the following result :

Lemma 4.8 (see [1]).

a) $\|A^{1l2}U|_{\varepsilon_\alpha,\sigma,\alpha} \leq \|A^{1l2}U|^{1-\varepsilon_\alpha} . |A^{1l2}U|^{\varepsilon_\alpha}_{\sigma,\alpha}$

b) $\|AU|_{\varepsilon_\alpha.\sigma,\alpha} \cdot \leq \|AU|^{1-\varepsilon_\alpha} . |AU|^{\varepsilon_\alpha}_{\sigma,\alpha} \cdot$

The proof is similar to the one given in [1].

Lemma 4.9 (see [1]). *Let* $U = (v, w) \in D\left(e^{\sigma A^{\frac{\alpha}{2}}} A\right)$, $\alpha \in]0, 1[$.

We have the following estimate :

$$\left|(B(U, U), AU)_{\sigma,\alpha}\right| \leq c_1 |A^{1l2}U|^{4/3}_{\sigma,\alpha} \cdot \left\|A^{\frac{1}{2}}U\right|^{1/6} . \|AU|^{1/6} . |AU|^{4/3}_{\sigma,\alpha} \cdot$$

Proof. By reporting (a) and (b) from lemma 4.8 into the inequality of the proposition 4.6, we get :

$$\left|(B(U, U), AU)_{\sigma,\alpha}\right| \leq c_1 |A^{1l2}U|^{1-\frac{\varepsilon_\alpha}{2}}_{\sigma,\alpha} \cdot \left\|A^{\frac{1}{2}}U\right|^{1+\frac{\varepsilon_\alpha}{2}} . \|AU|^{1+\frac{\varepsilon_\alpha}{2}} . |AU|^{1-\frac{\varepsilon_\alpha}{2}}_{\sigma,\alpha} .$$

For $\varepsilon_\alpha = \frac{2}{3}$ that is for $\alpha = \alpha_\star = \dfrac{\log(5/3)}{\log 2}$, we get the inequality stated in the lemma 4.9. □

We can now state our theorem of regularity.

Theorem 4.10. *Suppose that* $U(0) = G \in H \times \mathbb{L}^2(Q)$ *and for given* $\delta > 0$, $\alpha_0 \in]o, 1[$*,that* $F \in \mathbb{L}^\infty\left(\mathbb{R}_+, D\left(e^{\delta A^{\frac{\alpha_0}{2}}}\right)\right)$. *Then there exists a (weak) solution of problem (3.7) and an open set* Ω *in* $]0, +\infty[$ *whose complementary is of measure zero such that :*

a) $\forall \sigma > 0$, $0 < \alpha < \alpha_0$, *the function :*

$$\Omega \longrightarrow D\left(e^{\sigma A^{\frac{\alpha}{2}}}.A^{\frac{1}{2}}\right)$$
$$t \mapsto U(t)$$

be continuous.

b) if $\alpha_0 > \alpha_* = \dfrac{\ln(5/3)}{\ln 2}$, *we have :*

$$\int_0^T \frac{\left\|e^{\sigma A^{\frac{\alpha_*}{2}}}.AU(t)\right|^2}{\left(1 + \left\|e^{\sigma A^{\frac{\alpha_*}{2}}}.A^{\frac{1}{2}}U(t)\right|^2\right)^2} dt \leq K(1+T)$$

Where $K = K(v,\ Q,\ \|F\|,\ |U(0)|)$, *with* $\|F\| = \|F\|_{\mathbb{L}^\infty\left(\mathbb{R}_+,\ D\left(e^{\sigma\mathcal{A}\frac{\alpha_*}{2}}\right)\right)}$.

Proof. For $\mathcal{A}U \in D\left(e^{\sigma\mathcal{A}\frac{\alpha}{2}}\right)$ or equivalently $U \in D\left(e^{\sigma\mathcal{A}\frac{\alpha}{2}}\mathcal{A}\right)$, we have :

$$\left(\frac{dU}{dt},\ \mathcal{A}U\right)_{\sigma,\alpha} + v(\mathcal{A}U,\ \mathcal{A}U)_{\sigma,\alpha} = (F,\ \mathcal{A}U)_{\sigma,\alpha} - (BU,\ \mathcal{A}U)_{\sigma,\alpha} - (S^\circ(U,\ \mathcal{A}U))_{\sigma,\alpha}$$

which yields

$$\frac{1}{2}\frac{d}{dt}\left(\mathcal{A}^{\frac{1}{2}}U,\ \mathcal{A}^{\frac{1}{2}}U\right)_{\sigma,\alpha} + v|\mathcal{A}U|^2_{\sigma,\alpha} = (F,\ \mathcal{A}U)_{\sigma,\alpha} - (BU,\ \mathcal{A}U)_{\sigma,\alpha} - (S^\circ(U,\ \mathcal{A}U))_{\sigma,\alpha}$$

which gives :

$$\frac{1}{2}\frac{d}{dt}\left|\mathcal{A}^{\frac{1}{2}}U\right|^2_{\sigma,\alpha} + v|\mathcal{A}U|^2_{\sigma,\alpha} \le |F|_{\sigma,\alpha}\cdot|\mathcal{A}U|_{\sigma,\alpha} + \left|(BU,\ \mathcal{A}U)_{\sigma,\alpha}\right| + \left|(S^\circ(U,\ \mathcal{A}U))_{\sigma,\alpha}\right|$$

The propositions 4.6 and 4.7 give for $\varepsilon_{\alpha_*} = \dfrac{2}{3}, \alpha_0 > \alpha_*,\ F \in D\left(e^{\delta\mathcal{A}\frac{\alpha}{2}}\right),\ \delta > 0$:

$$\begin{aligned}\frac{1}{2}\frac{d}{dt}\left|\mathcal{A}^{\frac{1}{2}}U\right|^2_{\sigma,\alpha_*} + v|\mathcal{A}U|^2_{\sigma,\alpha_*} \le\ & |F|_{\sigma,\alpha_*}\cdot|\mathcal{A}U|_{\sigma,\alpha_*} \\ & + C_1\left|\mathcal{A}^{\frac{1}{2}}U\right|^{4/3}_{\sigma,\alpha_*}\cdot|\mathcal{A}U|^{4/3}_{\sigma,\alpha_*}\cdot\left\|\mathcal{A}^{1l2}U\right|^{1/6}\cdot\|\mathcal{A}U|^{1/6} \\ & + c_0\left|\mathcal{A}^{\frac{1}{2}}U\right|^2_{\sigma,\alpha_*} + \tfrac{\nu}{8}|\mathcal{A}U|^2_{\sigma,\alpha_*} + c_\nu.\left|\mathcal{A}^{\frac{1}{2}}U\right|^2_{\sigma,\alpha_*}\end{aligned}$$

We first have :

$$|F|_{\sigma,\alpha_*}\cdot|\mathcal{A}U|_{\sigma,\alpha_*} \le \frac{8}{v}|F|^2_{\sigma,\alpha_*} + \frac{v}{8}|\mathcal{A}U|^2_{\sigma,\alpha_*}$$

then by Young's inequality : $ab \le \varepsilon a^p + C_\varepsilon b^q,\ 1 < p < \infty,\ \varepsilon > 0$. By taking $\varepsilon = \dfrac{v}{8C_1}$, $p = 3/2$ and $q = 3$, we have:

$$C_1\left|\mathcal{A}^{\frac{1}{2}}U\right|^{4/3}_{\sigma,\alpha_*}\cdot|\mathcal{A}U|^{4/3}_{\sigma,\alpha_*}\cdot\left\|\mathcal{A}^{1l2}U\right|^{1l6}\cdot\|\mathcal{A}U|^{1l6}$$

$$\le \tfrac{v}{8}|\mathcal{A}U|^2_{\sigma,\alpha_*} + C_v\left|\mathcal{A}^{\frac{1}{2}}U\right|^4_{\sigma,\alpha_*}..\left\|\mathcal{A}^{\frac{1}{2}}U\right|^{1l2}.|\mathcal{A}U|^{\frac{1}{2}}$$

It follows :

$$\frac{d}{dt}\left|\mathcal{A}^{\frac{1}{2}}U\right|^2_{\sigma,\alpha*}+v\left|\mathcal{A}U\right|^2_{\sigma,\alpha*}$$

$$\leq \widetilde{C}_v\left(\left\{|F|^2_{\sigma,\alpha*}+\left[\left\|\mathcal{A}^{1l2}U\right|.\left\|\mathcal{A}U\right|^{\frac{1}{2}}\right]\right\}.\left(1+\left|\mathcal{A}^{\frac{1}{2}}U\right|^2_{\sigma,\alpha*}\right)^2+\left|\mathcal{A}^{\frac{1}{2}}U\right|^2_{\sigma,\alpha*}\right).$$

and then :

$$\frac{\frac{d}{dt}\left|\mathcal{A}^{\frac{1}{2}}U\right|^2_{\sigma,\alpha*}+v\left|\mathcal{A}U\right|^2_{\sigma,\alpha*}}{\left(1+\left|\mathcal{A}^{\frac{1}{2}}U\right|^2_{\sigma,\alpha*}\right)^2}\leq \widetilde{C}_v\left\{|F|^2_{\sigma,\alpha*}+\left[\left\|\mathcal{A}^{1l2}U\right|.\left\|\mathcal{A}U\right|\right]^{1l2}+1\right\}. \quad (4.14)$$

Let now Ω be the open of regularity $D\left(e^{\sigma\mathcal{A}^{\frac{\alpha*}{2}}}\mathcal{A}^{1l2}\right)$ given in the theorem 3.5, $\Omega = \bigcup_j]a_j,\ b_j[$,and $]a_j,\ b_j[$ interval of maximal regularity :

$$\left|\mathcal{A}^{\frac{1}{2}}U(t)\right|_{\sigma,\alpha*}\longrightarrow +\infty\ as\ t\longrightarrow b_j$$

If $\Lambda(t)$ denotes the first member of the inequality (4.14), we get by integration on $(a_j,\ b_j)$:

$$\int_{a_j}^{b_j}\Lambda(t)\,dt = v\int_{a_j}^{b_j}\frac{|\mathcal{A}U|^2_{\sigma,\alpha*}}{\left(1+\left|\mathcal{A}^{\frac{1}{2}}U\right|^2_{\sigma,\alpha*}\right)^2}dt+\frac{1}{\left(1+\left|\mathcal{A}^{\frac{1}{2}}U(a_j^-)\right|^2_{\sigma,\alpha*}\right)}$$

It follows :

$$\int_{a_j}^{b_j}\frac{|\mathcal{A}U|^2_{\sigma,\alpha*}}{\left(1+\left|\mathcal{A}^{\frac{1}{2}}U\right|^2_{\sigma,\alpha*}\right)^2}dt\leq C_\nu\left(\|F\|_{L^\infty}(b_j-a_j)+(b_j-a_j)+\int_{a_j}^{b_j}\left(\left\|\mathcal{A}^{1l2}U\right|.\|\mathcal{A}U|\right)^{1l2}dt\right)$$

It follows that for all positive T :

$$\int_0^T\frac{|\mathcal{A}U|^2_{\sigma,\alpha*}}{\left(1+\left|\mathcal{A}^{\frac{1}{2}}U\right|^2_{\sigma,\alpha*}\right)^2}dt\leq C_\nu\left(\|F\|_{L^\infty}.T+T+\int_0^T\left(\left\|\mathcal{A}^{1l2}U\right|.\|\mathcal{A}U|\right)^{1l2}dt\right)$$

Now by using Hölder's inequality with $p=\frac{4}{3}$ and $q=4$ in the second integral we get : .

$$\int_0^T\frac{|\mathcal{A}U|^2_{\sigma,\alpha*}}{\left(1+\left|\mathcal{A}^{\frac{1}{2}}U\right|^2_{\sigma,\alpha*}\right)^2}dt\leq T.C_\nu\|F\|_{L^\infty}+T+$$

$$\int_0^T\left(1+\left\|\mathcal{A}^{1l2}U\right|^2\right)^{3/4}\left\{\frac{\|\mathcal{A}U|^2}{\left(1+\|\mathcal{A}^{1l2}U|^2\right)}\right\}^{1l4}dt$$

$$\int_0^T \frac{|\mathcal{A}U|^2_{\sigma,\alpha_*}}{\left(1+\left|\mathcal{A}^{\frac{1}{2}}U\right|^2_{\sigma,\alpha_*}\right)^2}dt \leq T.C_\nu \|F\| + T$$

$$+\left\{\int_0^T \left(1+\left\|\mathcal{A}^{1/2}U\right|^2\right)dt\right\}^{3/4}\left\{\int_0^T \frac{\|\mathcal{A}U(t)|^2}{\left(1+\|\mathcal{A}^{1/2}U|^2\right)^2}dt\right\}^{1/4}$$

Now, by using the theorem 3.5 (W. Chen), we get the looked for estimation :

$$\int_0^T \frac{|\mathcal{A}U|^2_{\sigma,\alpha_*}}{\left(1+|\mathcal{A}^{1/2}U|^2_{\sigma,\alpha_*}\right)^2}dt \leq L_\nu(1+T)$$

where $L_\nu = L_\nu(\nu,\ Q,\ \|F\|_{L^\infty},\ U(0))$.

For the continuity, the proof is inspired from [5] and [3], and by using the properties of Ω. Remark first that :

for all $\sigma > 0$:

$$D\left(e^{\sigma\mathcal{A}^{\frac{\alpha}{2}}}\mathcal{A}^{1/2}\right) \subset D\left(\mathcal{A}^{1/2}\right).$$

Then thanks to theorem 3.5 we have the existence of a solution of problem (3.4) and the existence of an open Ω in $\mathbb{R}_+$ whose complementary has zero measure such that :

$$U : \Omega \to D\left(\mathcal{A}^{1/2}\right),\ continuous.$$

for all $t_0 \in \Omega$, $\left\|\mathcal{A}^{1/2}U(t_0)\right| < \infty$. The theorem 3.4 shows that there exists $T_0 > 0$, a regular solution

$$\begin{aligned}[t_0,\ T_0] &\longrightarrow H \times \mathbb{L}^2(Q)\\ t &\mapsto e^{\delta t\mathcal{A}^{\frac{\alpha_0}{2}}}.\mathcal{A}^{\frac{1}{2}}U(t),\quad \text{continuous.}\end{aligned}$$

As for all $\sigma > 0$, $\sigma_0 > 0$ and $0 < \alpha < \alpha_0$, we have :

$$D\left(e^{\delta t_0\mathcal{A}^{\frac{\alpha_0}{2}}}.\mathcal{A}^{\frac{1}{2}}\right) \hookrightarrow D\left(e^{\sigma\mathcal{A}^{\frac{\alpha}{2}}}.\mathcal{A}^{\frac{1}{2}}\right).$$

In particular, $\left\|e^{\sigma\mathcal{A}^{\frac{\alpha}{2}}}.\mathcal{A}^{\frac{1}{2}}U(t_0)\right| < \infty$ and thus for all t in Ω, $U(t) \in D\left(e^{\sigma\mathcal{A}^{\frac{\alpha}{2}}}.\mathcal{A}^{\frac{1}{2}}\right)$.
Considernow the following sets :

$$O = \left\{ t \in (0,\ +\infty) \,/\, U(t) \in D\left(e^{\sigma \mathcal{A}^{\frac{\alpha}{2}}}.\mathcal{A}^{\frac{1}{2}}\right) \right\}$$
$$\Sigma = (0,\ +\infty) \diagdown O.$$

Of course Ω writes as :

$$\Omega = \{ t \in (0,\ +\infty) \,/\, U(t) \text{ continuous in a neighborhood of } t \}$$
$$= \left\{ t \in (0,\ +\infty) \,/\, \text{there exists } \varepsilon > 0 : U \in \mathcal{C}\left[(t-\varepsilon,\ t+\varepsilon)\,;\ D\left(e^{\sigma \mathcal{A}^{\frac{\alpha}{2}}}.\mathcal{A}^{\frac{1}{2}}\right) \right] \right\}$$

Remark that Ω is open. Then Σ is of measure zero because of the initial condition of problem (3.4) since $U_0 \in H \times H$. Now if $t_0 \in O \setminus \Omega$, the theorem 3.2 page 22 of [7] shows that t_0 is the end of an interval of regularity $D\left(e^{\sigma \mathcal{A}^{\frac{\alpha}{2}}}.\mathcal{A}^{\frac{1}{2}}\right)$ that is to say that t_0 is an element of a connected component of O and thus $O \setminus \Omega$ is countable. It follows that $(0,\ +\infty)\setminus O$ is of measure zero. This ends the proof of the theorem .

□

References

[1] W. Chen, New a priori estimates in Gevrey class of regularity for weak solutions of three D. Navier-Stokes equations. *Diff. integral. Equations*, **7** (1994), 101-107.

[2] A. C. Eringen, Theory of micropolar fluids. *J. Math. Mecha.* **16** (1966), 1-18.

[3] C. Foias, Guillope and R. Temam, New a priori estimates for N. S. equations. *C.P.D.E.* **6** (1981), 329-359.

[4] C. Foias and R. Temam, Gevrey class regularity. *Journal of Functional Analysis*, **87/2** (1989).

[5] C. Foias and R. Temam, Some analytic and geometric properties of the solutions of N. S. equations. *J.M.P. Appl.* **58** (1979), 339-368.

[6] R. Temam: *Navier-Stokes equations: theory and numerical analysis.* North-Holland. 1979.

[7] R. Temam: *Navier-Stokes equations and nonlinear analysis.* N.S.F./C.BM.S. S.I.has.M. Philadelphia 1983.

In: Focus on Evolution Equations
Editor: Gaston M. N'Guerekata

ISBN: 978-1-60021-342-7

Chapter 7

EXISTENCE OF GENERALIZED SOLUTIONS FOR ORDINARY DIFFERENTIAL EQUATIONS IN BANACH SPACES

Thomas I. Seidman*
Department of Mathematics and Statistics,
University of Maryland Baltimore County,
Baltimore, MD, U.S.A.

Abstract

We introduce a hypothesis: that $f(\xi) \subset \overline{\text{hull}}\ \{\Phi(\xi)\}$ for a suitable set Φ of Lipshitzian functions and use a compactness result for parametrized sets of fixpoints to show that this is sufficient to ensure the existence of (generalized) solutions for the Banach space ordinary differential equation $\dot{x} = \mathbf{A}x + f(x)$.

1. Introduction

We will be considering the existence of solutions of ordinary differential equations in Banach spaces, taking these to have the nominal form

$$\dot{x} = \mathbf{A}x + f(x), \qquad\qquad x(0) = \hat{\xi}_0 \tag{1.1}$$

on some interval $[0, T]$. Here $x(\cdot)$ takes values in the Banach space $\mathcal{X}$ and $\mathbf{A}$ is the infinitesimal generator of a C_0 semigroup $\mathbf{S}(\cdot)$ of linear operators on $\mathcal{X}.f : \mathcal{X} \to \mathcal{X}$; note that no compactness assumption is to be imposed on $\mathbf{S}(t)$. We are indebted to [8] for an excellent survey of the extensive research on this problem; see also references there, especially [5].

For uniformly Lipschitzian $f : \mathcal{X} \to \mathcal{X}$, a standard formulation of 'solution' for (1.1) is in terms of the integral equation

$$x(t) = \mathbf{S}(t)\hat{\xi}_0 + \int_0^t \mathbf{S}(t-s) f(x(s))\, ds \qquad (0 \le t \le T), \tag{1.2}$$

*E-mail address: seidman@math.umbc.edu

where, through Banach's Contraction Mapping Principle (CMP), the uniform Lipschitz condition ensures existence.

When f is merely continuous, however, Peano [6] showed local existence for $\mathcal{X} = \mathbb{R}^n$, but Godunov [4] (following an example by Dieudonné [2]) showed that this fails for all infinite dimensional Banach spaces: there always exists continuous $f : \mathcal{X} \to \mathcal{X}$ such that (1.1) with $\mathbf{A} = 0$ has no local solution. Further, for discontinuous f a more general solution notion is appropriate: we modify (1.2) to define a 'solution' of (1.1) as a pair of $\mathcal{X}$-valued functions $[x, y]$ on $[0, T]$ — with $x(\cdot)$ absolutely continuous and $y(\cdot)$ locally integrable — such that

$$x(t) = \mathbf{S}(t)\hat{\xi}_0 + \int_0^t \mathbf{S}(t-s)y(s)\,ds \qquad (0 \leq t \leq T) \tag{1.3}$$

where, rather than merely having $y(s) = f(x(s))$ we extend the nonlinearity, following Fillipov [3], and require instead that

$$y(s) \in F_0(x(s)) \qquad (0 \leq s \leq T) \quad \text{with } F_0(\hat{\xi}) = \bigcap_{\varepsilon>0} F_\varepsilon(\hat{\xi}) \tag{1.4}$$

or, more precisely, that

$$\begin{gathered} y(\cdot) \in \mathcal{Y}_0(x(\cdot)) = \bigcap_{\varepsilon>0} \mathcal{Y}_\varepsilon(x) \qquad \text{with} \\ \mathcal{Y}_\varepsilon(x) = \left\{ y \in L^2([0,T] \to \mathcal{X}) : y(s) \in F_\varepsilon(x(s)) \text{ ae } s \in [0,T] \right\}. \end{gathered} \tag{1.5}$$

Noting that only the set-valued function $F_0(\cdot)$ is now relevant, we also permit $f(\cdot)$ to be set-valued — specifically permitting the possibility that $f(\xi) = \emptyset$ — and define

$$F_\varepsilon(\hat{\xi}) = F_\varepsilon(\hat{\xi}; f) = \overline{\text{hull}} \left\{ \zeta + \eta : \|\eta\| \leq \varepsilon,\ \zeta \in f(\xi),\ \|\xi - \hat{\xi}\| \leq \varepsilon \right\}. \tag{1.6}$$

Note that although we take the pair $[x, y]$ as defining a solution, the function y is uniquely determined by x; see Remark 7.

Remark 1. Fillipov, working in a finite dimensional context, actually included exclusion of nullsets in the definition (1.6). The point of this was to restrict consideration to (locally) 'typical' values of $f(\xi)$ and we can accomplish the same purpose, instead, by 'trimming' $f(\cdot)$ — omitting atypical values even if this would give an empty 'value', which our present formulation admits. Although we do not consider this here, a similar trimming might be relevant to treating invariant sets, considering the existence of solutions which remain in some specified subset of $\mathcal{X}$. ■

Especially in view of the known nonexistence examples for continuous $f(\cdot)$, our interest here lies in providing a suitable condition on f to ensure existence of a solution in the sense described above. To this end, we consider a set Φ of functions $\varphi : \mathcal{X} \to \mathcal{X}$ such that

$$\begin{array}{ll} i.] & \text{topologized by uniform convergence on compact sets,} \\ & \Phi \text{ is compact,} \\ ii.] & \Phi \text{ is convex as a subset of } C(\mathcal{X} \to \mathcal{X}), \\ iii.] & \text{each } \varphi \in \Phi \text{ is Lipschitzian with a uniform constant } L. \end{array} \tag{1.7}$$

For $\xi \in \mathcal{X}$ we write $\Phi(\xi)$ for $\{\varphi(\xi) : \varphi \in \Phi\}$ and note that (1.7) ensures that each such $\Phi(\xi)$ is a compact convex subset of $\mathcal{X}$. Our main result, then, is the following:

Theorem 2. *Let* $\mathbf{A}$ *be the infinitesimal generator of a* C_0 *semigroup* $\mathbf{S}(\cdot)$ *on the Banach space* $\mathcal{X}$ *and let* f *be a set-valued function on* $\mathcal{X}$. *Suppose there is a set* Φ, *satisfying* (1.7) *as above, such that*

$$\mathcal{D} = \left\{\xi : \hat{f}(\xi) = f(\xi) \cap \Phi(\xi) \neq \emptyset\right\} = \mathcal{D}(\hat{f}) \tag{1.8}$$

is dense in $\mathcal{X}$. *Then the ordinary differential equation* (1.1) *has a global solution* $x(\cdot)$ *in the sense of* (1.3), (1.5), (1.6).

Remark 3. The requirement that $\mathcal{D}$ be dense ensures that $F_0(\xi) \supset F_0(\xi; \hat{f})$ is non-empty for each ξ. It is worth noting at this point, however, that $\xi \mapsto [\mathbf{A}\xi + f(\xi)]$ may actually be undefined for every ξ — the dense domain $\mathcal{D}$ of (1.8) might be entirely disjoint from the dense domain of the infinitesimal generator $\mathbf{A}$. Thus, (1.1) cannot at all be interpreted pointwise, but only through the integrated form (1.3).

It is easy to see that in the finite dimensional case ($\mathcal{X} = \mathbb{R}^n$) the introduction of Φ is not at all a local constraint, but only a growth condition on the nonlinearity, irrelevant for local existence. It is in the infinite dimensional case that our hypotheses will provide a necessary compactness, e.g., excluding the examples of [2] and [4]. We note that these hypotheses are substantially weaker (locally) than assuming that each point $\hat{\xi} \in \mathcal{X}$ has a neighborhood $\mathcal{N} = \mathcal{N}(\hat{\xi})$ with $\bigcup_{\xi \in \mathcal{N}} f(\xi)$ precompact. Our hypotheses are, of course, weaker than assuming that $\bigcup_{\xi \in \mathcal{K}} f(\xi)$ is precompact for each compact set $\mathcal{K} \subset \mathcal{X}$, but that assumption would hold for any continuous f and so could not ensure existence in view of [2], [4].

We note that our argument will actually show the existence of (generalized) solutions for: $\dot{x} - \mathbf{A}x \in \hat{f}(x)$. Since the convexity of Φ, hence of each $\Phi(\xi)$, ensures that

$$F_\varepsilon(\xi; \hat{f}) \subset F_\varepsilon(\xi; f) \cap [\Phi(\xi) + (L+1)\mathcal{B}_*] \tag{1.9}$$

(with $\mathcal{B}_* = \{\xi \in \mathcal{X} : |\xi| \leq 1\}$), we see that our solutions of (1.1) are necessarily also solutions of the differential inclusion

$$\dot{x} - \mathbf{A}x \in \Phi(x). \tag{1.10}$$

Our compactness argument will then rely on properties of the solution set of (1.10). ■

We emphasize that our present concern is only with the *existence* of solutions, not with uniqueness. Standard examples in $\mathbb{R}$ and $\mathbb{R}^n$ show that additional hypotheses are generally needed to ensure uniqueness and this is, of course, also a major concern of semigroup theory, both linear and nonlinear. For more general concerns with possible nonuniqueness we refer, e.g., to [3] and also note [10], [1]. Nevertheless, we do have a modicum of well-posedness, for which see Remark 10.

2. Strategy

Our strategy for the proof of Theorem 2 will be the usual strategy for proving Peano's Theorem:

- Use the 'forward Euler' approach to construct a sequence of approximations $\{[x_N, y_N]\}$.
- Use compactness to show suitable convergence of some subsequence of these approximations to a limit $[\bar{x}, \bar{y}]$.
- Verify that this limit $[\bar{x}, \bar{y}]$ satisfies (1.3), (1.5), (1.6).

It is precisely the lack of local compactness in infinite dimensional spaces which leads to the counterexamples of [2] and [4]. The first and third steps of this argument are quite straightforward and the point of this paper is that our hypothesis (1.7), (1.8) can provide the compactness needed for the second step. Our principal tools for this will be a representation for y and a general topological result from [9]:

Theorem 4. *Let $(\mathcal{S}, d)$ be a complete metric space and $\mathcal{M}$ an arbitrary index set; for $m \in \mathcal{M}$, let $G(\cdot, m) : \mathcal{S} \to \mathcal{S}$ satisfy a Lipschitz condition:*

$$d(G(\sigma, m), G(\sigma', m)) \leq \vartheta\, d(\sigma, \sigma') \qquad (\sigma, \sigma' \in \mathcal{S}) \tag{2.11}$$

(for some $\vartheta = \vartheta_m < 1$) so there is a fixpoint $\sigma_m = G(\sigma_m, m)$ for each $m \in \mathcal{M}$. Now suppose that $\vartheta_m = \vartheta$ is independent of m and that the set

$$G(\mathcal{K}, \mathcal{M}) = \{G(\sigma, m) : \sigma \in \mathcal{K},\ m \in \mathcal{M}\}$$

has compact closure in $\mathcal{S}$ for every compact subset $\mathcal{K} \subset \mathcal{S}$. Then every sequence $(\sigma_{m(k)})$ of fixpoints contains a subsequence converging in $\mathcal{S}$.

When we come to apply Theorem 4, the index set $\mathcal{M}$ will have the form

$$\begin{aligned} \mathcal{M} &= L^2([0,T] \to \underline{M}) \qquad \text{where} \\ \underline{M} &= \{\text{Borel measures on } \Phi\} \subset [C(\Phi)]^*. \end{aligned} \tag{2.12}$$

Note that the measures in $[C(\Phi)]^*$ are signed measures while the 'Borel measures' $\mu(\cdot)$ in $\underline{M}$ above are positive with $\mu(\Phi) = 1$.

Lemma 5. *Let $x \in C([0,T] \to \mathcal{X})$ and $z \in \boldsymbol{\mathcal{Y}} = L^2([0,T] \to \mathcal{X})$ with $z(s) \in \Phi(x(s))$ for each $s \in [0,T]$. Then there is a measure-valued function $m \in \mathcal{M}$ such that*

$$z(s) = \int_\Phi \varphi(x(s))\, m(s, d\varphi) \qquad \textit{with } [s \to m(s,\cdot)] \in \mathcal{M}. \tag{2.13}$$

[Pointwise this is trivial and nonunique — by Choquet theory [7] one can take $m(s,\cdot)$ to be supported by the extreme points of Φ or, e.g., one might more simply take $m(s, A) = \{1 \text{ if } \hat{\varphi} \in A;\ \ 0 \text{ else}\}$ where the condition $z(s) \in \Phi(x(s))$ gives $z(s) = \hat{\varphi}(x(s))$ for some $\hat{\varphi} \in \Phi$. Thus (2.13) involves a selection and this can be taken measurable (so $m \in \mathcal{M}$) for measurable z.]

Lemma 6. *Suppose we have a weak-* convergent sequence* $m_j \stackrel{*}{\rightharpoonup} \bar{m}$ *in* $\mathcal{M} \subset [L^2([0,T] \to C(\Phi))]^*$ *and a strongly convergent sequence* $x_j \to \bar{x}$ *uniformly on* $[0,T]$*; correspondingly define* $y_j, \bar{y}$ *by* (2.13) *with* $x = x_j, \bar{x}$*. Then we have weak convergence* $y_j \rightharpoonup \bar{y}$ *in* $\boldsymbol{\mathcal{Y}}$.

PROOF: For any $\eta \in \boldsymbol{\mathcal{Y}}^* = L^2([0,T] \to \mathcal{X}^*)$ and y as in (2.13), we consider

$$\begin{aligned} \langle \eta, y \rangle_{\boldsymbol{\mathcal{Y}}} &= \int_0^T \langle \eta(s), y(s) \rangle_{\mathcal{X}}\, ds \\ &= \int_0^T \left\langle \eta(s), \int_\Phi \varphi(x(s))\, m(s, d\varphi) \right\rangle_{\mathcal{X}} ds \qquad = \langle \hat{\eta}, m \rangle_{\mathcal{M}} \\ &\quad \text{with } \hat{\eta}(s,\varphi) = \langle \eta(s), \varphi(x(s)) \rangle_{\mathcal{X}} \quad \text{for ae } s \in [0,T],\ \varphi \in \Phi, \end{aligned}$$

noting that: $\varphi \mapsto \varphi(\xi) : \Phi \to \mathcal{X}$ is continuous so $[\varphi \mapsto \hat{\eta}(s,\varphi)]$ is in $C(\Phi)$ for each s. Applying the above to $\langle \eta, y_j \rangle_{\boldsymbol{\mathcal{Y}}}$ and letting $m_j \stackrel{*}{\rightharpoonup} \bar{m}$, noting that the Lipschitz condition ensures that $\varphi(x_j(\cdot)) \to \varphi(\bar{x}(\cdot))$ uniformly for $\varphi \in \Phi$, we then get $\langle \eta, y_j \rangle_{\boldsymbol{\mathcal{Y}}} \to \langle \eta, \bar{y} \rangle_{\boldsymbol{\mathcal{Y}}}$. ■

3. Proof of Theorem 2

In this section we begin the proof of Theorem 2 following the strategy described in Section 2, noting that for global existence it is sufficient to prove existence on $[0,T]$ with $T > 0$ arbitrary. We now proceed sequentially with the steps of that strategy.

Our first step is to construct approximate solutions. Fix N, let $t_n = nT/N$ for $n = 0, 1, \ldots, N$, and then recursively define the approximate solution pair $[x_N, y_N]$ on the intervals $[t_n, t_{n+1}]$ as follows:
Take $\hat{\xi}_0$ as given in the initial condition and then, given $\hat{\xi}_n$ for $n < N$, arbitrarily choose some $\xi_n \in \mathcal{D}$ with $\|\xi_n - \hat{\xi}_n\| \leq T/N$ — possible as we have assumed $\mathcal{D} = \mathcal{D}(f, \Phi)$ is dense in $\mathcal{X}$. We can then arbitrarily choose $\zeta_n \in \hat{f}(\xi_n) \neq \emptyset$ and define

$$y_N(s) \equiv \zeta_n \qquad x_N(s) = \mathbf{S}(s - t_n)\hat{\xi}_n + \int_{t_n}^s \mathbf{S}(s - r)\, \zeta_n\, dr \tag{3.14}$$

for $s \in (t_n, t_{n+1}]$. Finally, we set $\hat{\xi}_{n+1} = x_N(t_{n+1})$. Note that this construction gives the integral relation (1.3) for $[x_N, y_N]$.

The next step is the compactness argument. This will be the longest part of the proof and, using Theorem 4, will rely entirely on (1.3) and the fact that for $nT/N < s \leq (n+1)T/N$ we have

$$y_N(s) = \zeta_n \in \Phi(\xi_n) \qquad \text{with } \|\xi_n - x_N(nT/N)\| \leq T/N. \tag{3.15}$$

[We focus on the specific approximate solutions $\{[x_N, y_N]\}$ as constructed above, but for this step it might be noted that we are looking at these primarily as approximate solutions of (1.10).]

Note first that $\zeta_n \in \Phi(\xi_n)$ just means that $\zeta_n = \varphi_n(\xi_n)$ for some $\varphi_n \in \Phi$ so, much as in Lemma 5, there is a measure-valued function $m_N : [0,T] \to \underline{M}$ given by

$$\begin{aligned} m_N(s,\cdot) &= \mu_n(\cdot) \quad \text{for } nT/N < s \le (n+1)T/N \\ &\text{where } \int_\Phi \varphi\, \mu_n(d\varphi) = \varphi_n \end{aligned} \tag{3.16}$$

from which we have

$$\begin{aligned} y_N(s) &= \zeta_n = \int_\Phi \varphi(\xi_n)\, m_N(s,d\varphi) \\ &= \int_\Phi \varphi(x_N(s))\, m_N(s,d\varphi) + \eta_N(s) \qquad \text{where} \\ \eta_N(s) &= \int_\Phi [\varphi(\xi_n) - \varphi(x_N(s))]\; m_N(s,d\varphi). \end{aligned} \tag{3.17}$$

We now set $\boldsymbol{\mathcal{X}} = C([0,T] \to \mathcal{X})$ and define $G(\cdot,m) : \boldsymbol{\mathcal{X}} \to \boldsymbol{\mathcal{X}}$ by

$$[G(x(\cdot),m)]\,(t) = \mathbf{S}(t)\hat{\xi}_0 + \int_0^t \int_\Phi \mathbf{S}(t-s)\, \varphi(x(s))\, m(s,d\varphi)\, ds, \tag{3.18}$$

parametrized by $m \in \mathcal{M}$ — so $\mathcal{M}$ serves only as an index set at present. Note that (3.14) then gives

$$x_N(t) = [G(x_N, m_N)]\,(t) + \int_0^t \eta_N(s)\, ds, \tag{3.19}$$

i.e., x_N is a fixpoint of $[G(\cdot,m_N) + e_N] : \boldsymbol{\mathcal{X}} \to \boldsymbol{\mathcal{X}}$ where $e_N = \int_0^t \eta_N$.

At this point we use a standard trick: selecting for $\boldsymbol{\mathcal{X}}$ the exponentially weighted norm

$$\|x\|_{\boldsymbol{\mathcal{X}}} = \max\left\{e^{-2\gamma L t}\, \|x(t)\|_{\mathcal{X}} : t \in [0,T]\right\} \tag{3.20}$$

where γ is a bound for $\|\mathbf{S}(t)\|$ on $[0,T]$; clearly this is equivalent to the usual (unweighted) max norm for $\boldsymbol{\mathcal{X}} = C([0,T] \to \mathcal{X})$. Now, for $0 \le t \le T$,

$$\begin{aligned} &e^{-2\gamma L t}\|\,[G(x_1,m) - G(x_2,m)]\,(t)\|_{\mathcal{X}} \\ &\quad = e^{-2\gamma L t} \left\| \int_0^t \mathbf{S}(t-s) \int_\Phi [\varphi(x_1(s)) - \varphi(x_2(s))]\; m(s,d\varphi)\, ds \right\|_{\mathcal{X}} \\ &\quad \le e^{-2\gamma L t} \int_0^t \|\mathbf{S}(t-s)\| \int_\Phi \|\varphi(x_1(s)) - \varphi(x_2(s))\|_{\mathcal{X}}\; m(s,d\varphi)\, ds \\ &\quad \le e^{-2\gamma L t} \int_0^t \gamma \int_\Phi L\, \|x_1(s) - x_2(s)\|_{\mathcal{X}}\; m(s,d\varphi)\, ds \\ &\quad = \gamma L \int_0^t e^{-2\gamma L(t-s)} \left[e^{-2\gamma L s}\, \|x_1(s) - x_2(s)\|_{\mathcal{X}}\right] ds \\ &\quad \le \gamma L \int_0^t e^{-2\gamma L(t-s)}\, ds\; \|x_1 - x_2\|_{\boldsymbol{\mathcal{X}}} \qquad \le \tfrac{1}{2}\, \|x_1 - x_2\|_{\boldsymbol{\mathcal{X}}} \end{aligned}$$

Thus, using the norm (3.20), each $G(\cdot,m)$ is contractive on $\boldsymbol{\mathcal{X}}$ with the uniform contraction constant $\vartheta = 1/2$. As in Theorem 4, we will denote the unique fixpoint of $G(\cdot,m)$ by σ_m for each $m \in \mathcal{M}$.

We now complete verification of the hypotheses of Theorem 4, i.e., we proceed to show that $\overline{G(\mathcal{K},\mathcal{M})}$ is compact in $\boldsymbol{\mathcal{X}}$ for any compact set $\mathcal{K} \subset \boldsymbol{\mathcal{X}}$. To this end, note that continuity of the evaluation map: $[x,t] \mapsto x(t)$ shows that the image K_1 of $\mathcal{K} \times [0,T]$ is compact. Similarly, the image K_2 of $\Phi \times K_1$ under $[\varphi,\xi] \mapsto \varphi(\xi)$ is also compact in $\mathcal{X}$ and the set of possible integrand values $K_3 = \{\mathbf{S}(\tau)\zeta : 0 \le \tau \le T,\ \zeta \in K_2\}$ is compact. Also, $K_0 = \{\mathbf{S}(t)\hat{\xi}_0 : t \in [0,T]\}$ is compact. The definition of $G(\cdot,m)$ then ensures that each $z \in G(\mathcal{K},\mathcal{M})$ takes its values in the compact set

$$K = K(\mathcal{K}) = K_0 + [0,T]\,\overline{\text{hull}}\,\{0, K_3\} \subset \mathcal{X}. \tag{3.21}$$

Now let

$$\begin{aligned} \beta(\mathcal{K}) &= \max\|\zeta\|_{\mathcal{X}} : \zeta \in K(\mathcal{K})\} \\ \omega(h,\mathcal{K}) &= \max\{\|\mathbf{S}(\tau)\xi - \xi\| : \xi \in K(\mathcal{K}),\ \tau \in [0,h]\}. \end{aligned}$$

By the compactness of $K = K(\mathcal{K})$, we have $\beta(\mathcal{K}) < \infty$ and $\omega(h,\mathcal{K}) \to 0$ as $h \to 0$, since $\mathbf{S}$ is a C_0 semigroup and $K(\mathcal{K})$ is compact. Next, we note that, for any $z(\cdot) \in G(\mathcal{K},\mathcal{M})$ and any $0 \le t < t' \le T$, one has from (1.3) that

$$z(t') = \mathbf{S}(t'-t)z(t) + \int_t^{t'} \mathbf{S}(t'-s)\zeta(s)\,ds$$

with $\zeta(s) \in K$. It follows that for $|t'-t| \le h$ one has

$$\|z(t+h) - z(t)\|_{\mathcal{X}} \le \omega(h,K) + h\gamma\beta(K)$$

so $G(\mathcal{K},\mathcal{M})$ is uniformly equicontinuous. The desired compactness of $\overline{G(\mathcal{K},\mathcal{M})}$ now follows from the (generalized) Arzelà-Ascoli Theorem.

We can now apply Theorem 4 to show that the 'fixpoint set' $\mathcal{K}_* = \{\sigma_m : m \in \mathcal{M}\}$ is precompact in $\boldsymbol{\mathcal{X}}$. In particular, for the sequence (σ_{m_N}), there is a convergent subsequence, with $N = N(j) \to \infty$. Abusing notation slightly, we write σ_j for $\sigma_{m_{N(j)}}$ so we have $\sigma_j \to \bar{x}$ uniformly on $[0,T]$.

4. Proof of Theorem 2 (continued)

In the previous section we constructed a sequence of approximate solutions (x_N) which were fixpoints of maps $[G(\cdot,m_N) + e_N] : \boldsymbol{\mathcal{X}} \to \boldsymbol{\mathcal{X}}$. Applying Theorem 4, we showed subsequential convergence $\sigma_j \to \bar{x}$ for the fixpoints $\sigma_j = G\left(\sigma_j, m_{N(j)}\right)$.

Our next task is to show that also $x_j \to \bar{x}$ (again abusing notation slightly in writing x_j for $x_{N(j)}$). Returning to (3.14) and (3.15) and taking $N = N(j)$, we now write

$$\begin{aligned} x_j(t) &= \mathbf{S}(t)\hat{\xi}_0 + \int_0^t \int_\Phi \mathbf{S}(t-s)\varphi(\hat{x}_j(s))\,m_N(s,d\varphi)\,ds \\ \text{where} \quad & \hat{x}_j(s) = \zeta_n \qquad \text{for } s \in [nT/N, (n+1)T/N] \\ \text{with} \quad & \|\zeta_n - x_j(nT/N)\|_{\mathcal{X}} \le T/N, \\ \sigma_j(t) &= \mathbf{S}(t)\hat{\xi}_0 + \int_0^t \int_\Phi \mathbf{S}(t-s)\varphi(\sigma_j(s))\,m_N(s,d\varphi)\,ds. \end{aligned} \tag{4.22}$$

For $s \in [nT/N, (n+1)T/N]$ we have

$$\begin{aligned}
&\|\hat{x}_j(s) - \sigma_j(s)\|_{\mathcal{X}} \\
&\quad \le T/N + \|x_j(nT/N) - \sigma_j(nT/N)\|_{\mathcal{X}} + \|\sigma_j(nT/N) - \sigma_j(s)\|_{\mathcal{X}} \\
&\quad \le T/N + \rho_j(nT/N) + \hat{\omega}(T/N) \qquad \le [T/N + \hat{\omega}(T/N)] + \rho_j(s)
\end{aligned}$$

where

$$\rho_j(s) = \max\{\|x_j(t) - \sigma_j(t)\|_{\mathcal{X}} : 0 \le t \le s\} \tag{4.23}$$

(making $\rho_j(\cdot)$ nondecreasing, so $\rho_j(nT/N) \le \rho_j(s)$) and

$$\hat{\omega}(h) = \max\left\{\|\sigma(t') - \sigma(t)\|_{\mathcal{X}} : |t' - t| \le h,\, \sigma \in \mathcal{K}_*\right\}. \tag{4.24}$$

Note that the compactness of $\overline{\mathcal{K}_*}$ in $\boldsymbol{\mathcal{X}}$ ensures that $\hat{\omega}(h) = \hat{\omega}(h, \mathcal{K})$ is well-defined with $\hat{\omega}(h) \to 0$ as $h \to 0$. We now proceed to estimate ρ_j:
For any $t \in [0, T]$ we have

$$\begin{aligned}
&\|x_j(t) - \sigma_j(t)\|_{\mathcal{X}} \\
&\quad = \left\| \int_0^t \int_\Phi \mathbf{S}(t-s)\,[\varphi(\hat{x}_j(s)) - \varphi(\sigma_j(s))]\, m_N(s, d\varphi)\, ds \right\|_{\mathcal{X}} \\
&\quad \le \gamma L \int_0^t \|\hat{x}_j(s) - \sigma_j(s)\|_{\mathcal{X}}\, ds
\end{aligned}$$

so, again noting that $\rho_j(\cdot)$ is nondecreasing, we have

$$\rho_j(t) \le \gamma L t\,[h + \hat{\omega}(h)] + \gamma L \int_0^t \rho_j(s)\, ds$$

with $h = h_j = T/N(j) \to 0$. Applying the Gronwall Inequality to this, we have $\|x_j - \sigma_j\|_{\boldsymbol{\mathcal{X}}} \le C[h + \hat{\omega}(h)]$ with a constant C depending only on $\mathcal{K}_*, T, \gamma, L$. Since $\sigma_j \to \bar{x}$ for some $\bar{x} \in \boldsymbol{\mathcal{X}}$, this shows that we also have $x_j \to \bar{x}$ uniformly on $[0, T]$.

We turn now to the sequence (y_j) which, corresponding to (4.22), is given by

$$y_j(s) = \int_\Phi \varphi(\hat{x}_j(s))\, m_{N(j)}(s, d\varphi), \tag{4.25}$$

i.e., y_j is given by (2.13) with $m = m_{N(j)}$ and $x = \hat{x}_j$; note that we have shown uniform convergence $\hat{x}_j \to \bar{x}$ on $[0, T]$. Clearly $\mathcal{M}$ is closed and, since $\mu \in \underline{M}$ gives $0 \le \mu(\cdot) \le 1$, is bounded as a subset of the dual space

$$L^2([0, T] \to [C(\Phi)]^*) = \left[L^2([0, T] \to C(\Phi))\right]^*.$$

Hence, by Alaoglu's Theorem, $\mathcal{M}$ is weak-* compact and (again extracting a further subsequence, if necessary) we have $m_{N(j)} \overset{*}{\rightharpoonup} \bar{m}$ for some $\bar{m} \in \mathcal{M}$. We may then apply Lemma 6 to have weak convergence $y_j \rightharpoonup \bar{y}$ in $\boldsymbol{\mathcal{Y}} = L^2([0, T] \to \mathcal{X})$ with $[\bar{x}, \bar{y}]$ also related by (2.13) using $\bar{m}$. Given any $\xi \in \mathcal{X}^*$ and $t \in [0, T]$ we set $\eta(s) = [\mathbf{S}(t-s)]^*\xi$ on $[0, t]$ and $\eta(s) = 0$ for $s > t$. Then $\eta \in \boldsymbol{\mathcal{Y}}^*$ and

$$\langle \xi, x_j(t) \rangle_{\mathcal{X}} = \langle \xi, \mathbf{S}(t)\hat{\xi}_0 \rangle_{\mathcal{X}} + \langle \eta, y_j \rangle_{\boldsymbol{\mathcal{Y}}}$$

since our construction gave (1.3) for $[x_j, y_j]$, as noted following (3.14). Since we have $x_j(t) \to \bar{x}(t)$ in $\mathcal{X}$ and $y_j \rightharpoonup \bar{y}$ in $\mathcal{Y}$ as $j \to \infty$, we also have (1.3) for $[\bar{x}, \bar{y}]$ in the limit.

To complete the proof of Theorem 2, we now verify (1.5) for $\bar{y}$. Note that each $\mathcal{Y}_\varepsilon$ (for $\varepsilon > 0$) is closed and convex, hence weakly closed in $\mathcal{Y}$, with $\mathcal{Y}_\varepsilon \subset \mathcal{Y}_{\varepsilon'}$ for $0 < \varepsilon < \varepsilon'$. Our construction selected

$$y_j(s) \in \hat{f}(\hat{x}_j(s)) \subset f(\hat{x}_j(s))$$

so $y_j(s) \in F_\varepsilon(\bar{x}(s))$ whenever $\|\hat{x}_j(s) - \bar{x}(s)\|_{\mathcal{X}} \leq \varepsilon$. Since we have already noted that $\hat{x}_j \to \bar{x}$ uniformly on $[0, T]$, we have (for every $\varepsilon > 0$) $y_j \in \mathcal{Y}_\varepsilon$ for large enough j, depending on ε. The weak convergence $y_j \rightharpoonup \bar{y}$ then also ensures that $\bar{y} \in \mathcal{Y}_\varepsilon$ in the limit. This, for each $\varepsilon > 0$, gives (1.5).

We have now shown that $[\bar{x}, \bar{y}]$ is a solution of (1.1) in the sense we have determined and so have completed the proof of Theorem 2. ∎

5. Further Remarks

Remark 7. It is interesting to note that the construction used in the proof of Theorem 2 gives all possible solutions in the sense of (1.3), (1.5), (1.6). I.e., if we are given any solution pair $[\bar{x}, \bar{y}]$ we can find a sequence of approximations $[x^j, y^j]$ converging suitably to $[\bar{x}, \bar{y}]$ with x^j obtained from y^j by (1.3) and y^j 'admissible' — piecewise constant and obtained from x^j by selections from $\hat{f}(x^j(\cdot))$ as at the beginning of the proof of Theorem 2.

Our first observation is that $\bar{y}$ takes its values in the compact set $K_2 = \Phi(K_1)$ where $K_1 = \{\bar{x}(t) : t \in [0, T]\}$ is compact as $\bar{x}$ is continuous. Further, if $[\bar{x}, \bar{y}_1]$ and $[\bar{x}, \bar{y}_2]$ were solutions, then, setting $y = \bar{y}_1 - \bar{y}_2$, we have $\int_0^t \mathbf{S}(t-s)y(s)\,ds = 0$ for all $t \in [0, T]$ so $\int_{t-\varepsilon}^t \mathbf{S}(t-s)y(s)\,ds = 0$ for all $0 < t \leq T$ and small $\varepsilon > 0$ and, since $\mathbf{S}(\tau)\eta \to \eta$ uniformly in $\eta = y(s) \in K_2$, it follows that $y(s) \equiv 0$ — i.e., the component $\bar{y}$ of a solution pair is uniquely determined by $\bar{x}$.

If we can find admissible (y^j) with values approximatly in K_2, then an argument much like that in the proof of Theorem 2 shows that we have subsequential convergence $x^{j(k)} \to \hat{x}$ uniformly in $\mathcal{X}$-norm on $[0, T]$ so if we can show $\hat{x} = \bar{x}$ we have the desired $x^j \to \bar{x}$. To this end it is sufficient to know that

$$\int_0^T \langle \eta_t, y^j \rangle\,ds \quad \longrightarrow \quad \int_0^T \langle \eta_t, \bar{y} \rangle\,ds \tag{5.26}$$

for each η_t of the form: $\eta_t(s) = \begin{cases} \mathbf{S}^*(t-s)\,\bar{\eta} & \text{if } s \leq t \\ 0 & \text{else} \end{cases}$ where we need only take t from a dense set in $[0, T]$ and $\bar{\eta}$ from a set in $\mathcal{X}^*$ which separates points in the compact set $K(\{\bar{x}\})$ as in (3.21), so these may also be taken countable. To see this sufficiency, we note that we then have $\langle \bar{\eta}, x^j(t) \rangle \to \langle \bar{\eta}, \bar{x}(t) \rangle$ — and, of course, $\langle \bar{\eta}, x^j(t) \rangle \to \langle \bar{\eta}, \hat{x}(t) \rangle$ by the uniform strong convergence $x^j \to \hat{x}$ — so $< \bar{\eta}, \hat{x}(t) \rangle =< \bar{\eta}, \bar{x}(t) \rangle$ for each such $t, \bar{\eta}$, whence $\hat{x} \equiv \bar{x}$.

It only remains to show existence of admissible (y^j) satisfying (5.26). It is quite standard to approximate (say, in L^2-norm) by piecewise constant functions $(\bar{y}^j)$ with values in $\bar{y}^j(s) \in F_{\varepsilon_j}(\bar{x}(s)) \subset F_{\varepsilon'_j}(x^j(s))$ with $\varepsilon_j, \varepsilon'_j \to 0$. If $\hat{y}$ is one of these constant values (on some small interval) so $\hat{y} \in \overline{\text{hull}}\,\{\text{selectable values}\}$, then we can approximate $\hat{y}$ in $\mathcal{X}$ by a finite convex combination of such selectable values and so, by a standard argument, use a much finer partition to approximate the constant function $\hat{y}$ (on its interval) in the sense of the countable number of considerations (5.26) by a piecewise constant function using the selectable values. It is then the function we get by collecting these refinements for the approximation $\bar{y}^j$ on each of its subintervals which we take as y^j. ■

Remark 8. It is clear from the proof of Theorem 2 that we also have a local version: if the hypothesized set Φ consists of functions defined and satisfies (1.7) only for ξ in some neighborhood of the initial $\hat{\xi}_0$ and if $\mathcal{D}$ of (1.8) is dense in that neighborhood, then we have local existence of a solution, existence on an interval $[0, T]$ for some $T > 0$. [To see this we need only note that the approximate solutions cannot escape the neighborhood for an estimable interval $[0, T]$ and then the proof goes through without further modification.] If we supplement this local consideration by a growth consideration, then we again have global existence:

Theorem 9. *Let* $\mathbf{A}$ *be the infinitesimal generator of a* C_0 *semigroup* $\mathbf{S}(\cdot)$ *on the Banach space* $\mathcal{X}$ *and let* f *be a set-valued function on* $\mathcal{X}$. *Suppose there is a continuous nondecreasing function* $\gamma : [0, \infty) \to (0, \infty)$ *with* $\int_0^\infty [1/\gamma(r)]\, dr = \infty$ *and, for each* $\hat{\xi} \in \mathcal{X}$, *suppose there is a neighborhood* $\mathcal{N} = \mathcal{N}(\hat{\xi})$ *and a set* $\Phi = \Phi_{\mathcal{N}}$ *of functions* $\varphi : \mathcal{N} \to \mathcal{X}$ *such that*

$$\begin{array}{ll} i.] & \text{topologized by uniform convergence on compact subsets of } \mathcal{N}, \\ & \quad \Phi \text{ is compact}, \\ ii.] & \Phi \text{ is convex as a subset of } C(\mathcal{N} \to \mathcal{X}), \\ iii.] & \text{each } \varphi \in \Phi \text{ is Lipschitzian with a fixed constant } L = L(\mathcal{N}), \\ iv.] & \text{for each } \varphi \in \Phi \text{ and } \xi \in \mathcal{N} \text{ we have } |\varphi(\xi)| \le \gamma(|\xi|). \end{array} \tag{5.27}$$

Suppose further that each

$$\mathcal{D}_{\mathcal{N}} = \{\xi \in \mathcal{N} : f(\xi) \cap \Phi_{\mathcal{N}}(\xi) \neq \emptyset\} \tag{5.28}$$

is dense in $\mathcal{N} = \mathcal{N}(\hat{\xi})$. *Then the ordinary differential equation* (1.1) *has a global solution* $x(\cdot)$ *in the sense of* (1.3), (1.5), (1.6).

PROOF: This is the standard Zorn's Lemma argument for a solution with maximal interval of existence, considering the family of triples $[x, y, T]$ where $T > 0$ and $[x, y]$ is a solution pair on $[0, T)$ in the sense of (1.3), (1.5), (1.6), partially ordering this family by setting $[x, y, T] \prec [\hat{x}, \hat{y}, \hat{T}]$ if $0 < T < \hat{T}$ and x, y are the restrictions to $[0, T)$ of $\hat{x}, \hat{y}$. We need only show that maximality is impossible for $T < \infty$. To see this, note that if $x(T)$ is well-defined, then re-initializing the autonomous equation (1.1) to start at $\hat{\xi}_0 = x(T)$, one can continue $x(\cdot)$ past T by local existence — noting that having (1.3) on $[0, T]$ and then

$$x(t) = \mathbf{S}(t - T)x(T) + \int_T^t \mathbf{S}(t - s)y(s)\, ds \qquad (0 < T \le t \le T') \tag{5.29}$$

is just equivalent to having (1.3) on $[0, T']$.

On the other hand, our considerations give $|y(s)| \leq \gamma(|x(s)|)$ so, noting the existence of an exponential bound $\|\mathbf{S}(t)\| \leq Me^{\omega t}$ for the semigroup, we have

$$|x(t)| \leq Me^{\omega t}|\hat{\xi}_0| + \int_0^t Me^{\omega(t-s)}\gamma(|x(s)|)\, ds$$

from which, under our considerations on γ, we can obtain an *a priori* bound on $|x(\cdot)|$ for arbitrarily large t and so a bound on $|y(\cdot)|$ uniformly on $[0, T)$. If we now define $\hat{y} = y$ on $[0, T)$ with $\hat{y}(s) = 0$ for $s \geq T$, then $\hat{y}$ is bounded and the equation

$$\dot{x} = \mathbf{A}x + \hat{y}, \qquad x(0) = \hat{\xi}_0$$

has a continuous mild solution $\hat{x}$, coinciding with x on $[0, T)$ and continuous for all t. Thus x extends to the closed interval $[0, T]$, satisfying (1.3) at $t = T$ as well so the local extendability noted above shows that, as desired, the maximal solution given by Zorn's Lemma must be a global solution. ■

Remark 10. Although solutions may be nonunique, there is a certain degree of well-posedness for the solution sense we are using: the solution set is upper semicontinuous (usc) in its dependence on its data — both the initial condition $\hat{\xi}_0$ and the trimmed nonlinearity $\hat{f}$. More precisely, we suppose a sequence of problems as in Theorem 2 with $\hat{\xi}_0^j \to \hat{\xi}_0$ and with the upper semi-convergence property:

$$\text{If } \zeta \in f^j(\hat{\xi}), \text{ then } \|\bar{\zeta} - \zeta\| \leq \varepsilon_j \text{ for some } \bar{\zeta} \in F_0(\hat{\xi}; f) \tag{5.30}$$

with $\varepsilon_j \to 0$. We then claim that:

- If $([x^j, y^j])$ is any corresponding sequence of solutions, then there is a subsequence converging suitably to a pair $[\bar{x}, \bar{y}]$.
- If $[x^j, y^j]$ is a sequence of solutions with $x^j \to \bar{x}$ (uniformly on $[0, T]$), then $[\bar{x}, \bar{y}]$ is a solution of the limit problem (for a suitable $\bar{y}$).

To see the first assertion, we note that $([x^j, y^j])$ is a sequence of solutions of (1.10) with initial data $\hat{\xi}_0^j \to \hat{\xi}_0$ so essentially the same argument as in the proof of Theorem 2 shows that there is a subsequence with $x^{j(k)} \to \bar{x}$ in $C([0, T] \to \mathcal{X})$ and with $y^{j(k)} \rightharpoonup \bar{y}$ weakly in $\boldsymbol{\mathcal{Y}} = L^2([0, T] \to \mathcal{X})$. For the second assertion we note that (5.30) together with $\|x^j - \bar{x}\| \leq \hat{\varepsilon}_j$ give

$$y^j(s) \in F_0(x^j(s); f^j) \subset F_{\max\{\varepsilon_j, \hat{\varepsilon}_j\}}(x(s); f)$$

and again essentially the same argument as in the proof of Theorem 2 shows that (for a subsequence) $y^{j(k)} \rightharpoonup \bar{y}$ weakly in $\boldsymbol{\mathcal{Y}} = L^2([0, T] \to \mathcal{X})$ for some $\bar{y} \in \boldsymbol{\mathcal{Y}}$ and that $[\bar{x}, \bar{y}]$ is a solution of the limit problem. [Since $\bar{y}$ is uniquely determined by $\bar{x}$, the Subsubsequence Lemma then permits us to ignore the extraction of subsequences here.] ■

References

[1] J.C. Alexander and T.I. Seidman, *Sliding modes in intersecting switching surfaces, I: blending, Houston J. Math.* **24** (1998), pp. 545–569.

[2] J. Dieudonné, *Deux exemples singuliers d'équations différentielles, Acta Sci. Math.* (Szeged) **12B** (1950), pp. 38–40.

[3] A.F. Filippov, *Differential Equations with Discontinuous Righthand Sides,* Nauka, Moscow, 1985 [*transl.* Kluwer, Dordrecht, 1988].

[4] A.N. Godunov, *Peano's theorem in Banach spaces* (transl.), *Func. Anal. and Applic.* **9** (1975), pp. 53–55.

[5] O. Hájek, *Discontinuous differential equations, I and II, J. Diff. Eqns.* **32** (1979), pp. 149–185.

[6] G. Peano, *Sull'integrabilita delle equazioni differenziali di primo ordine, Accad. Sci. Torino* **21** (1885), pp. 853–857.

[7] Robert R. Phelps, *Lectures on Choquet's Theorem, 2nd ed.* (Lecture Notes in Mathematics #1757), Springer-Verlag, Berlin, 2001.

[8] E. Schecter, *A survey of local existence theories for abstract nonlinear initial value problems,* in *Nonlinear Semigroups, PDE, Attractors* (LNM #1394; T.E. Gill, W.W. Zachary, eds.), Springer-Verlag, New York, 1989, pp. 136–184.

[9] T.I. Seidman, *Two compactness lemmas,* in *Nonlinear Semigroups, PDE, Attractors* (LNM #1248; T.E. Gill, W.W. Zachary, eds.), Springer-Verlag, New York, 1987, pp. 162–168.

[10] T.I. Seidman, *The residue of model reduction,* in *Hybrid Systems III. Verification and Control,* (LNCS #1066; R. Alur, T.A. Henzinger, E.D. Sontag, eds.) pp. 201–207, Springer-Verlag, Berlin (1996).

[11] V.I. Utkin, *Sliding Models and their Application in Variable Structure Systems*, Mir, Moscow, 1978.

In: Focus on Evolution Equations
Editor: Gaston M. N'Guerekata
ISBN: 978-1-60021-342-7

Chapter 8

On the Existence of Almost Periodic, Periodic and Quasi-periodic Solutions of Neutral Differential Equations with Piecewise Constant Arguments

Tran Tat Dat*
Department of Mathematics, Hanoi University of Science,
334 Nguyen Trai, Hanoi, Vietnam

Abstract

This note is concerned with the almost periodicity, periodicity, and quasi periodicity of bounded solutions and a so-called Massera criterion for the existence of periodic solution of neutral differential equation with piecewise constant argument.

Keywords and phrases: Neutral differential equation with piecewise constant argument, Massera type criterion, spectrum of a function, almost periodic solution, periodic solution and quasi periodic solution.

1. Introduction

The differential equations with piecewise constant arguments (EPCAs, for short) are subjects which are only studied recently. These equations have the structure of continuous dynamical systems in intervals of unit length. Continuity of a solution at a point joining any two consecutive intervals implies a recursion relation for the values of the solution at such points. Therefore, they combine the properties of differential equations and difference equations. These equations are thus similar in structure to those found in certain sequential-continuous models of disease dynamics as treated by Busenberg and Cooke since 1982 (see [3]). The first contribution is due to Cooke and Wiener in 1984 see [5] and Shah and Wiener in 1983 (see [28]). Cooke and Wiener studied in [5, 6] the existence and uniqueness of solution and of its backward continuation on $(-\infty, 0]$ and investigated asymptotic stability of

*E-mail address: tatdat4382@yahoo.com

the trivial solution for equation $\dot{x}(t) = Ax(t) + \sum_{k=-N}^{N} A_k x([t+k]) + f(t), \quad x(t) \in \mathbb{R}$. The results about oscillation properties can be found in [30, 31, 32] and references therewith. The existence of periodic solutions has been studied in [7, 29, 30, 32] and references therewith. In 1991, Cooke and Wiener gave a survey paper, (see [7]), of described the results (before 1991) in the area of differential equations with piecewise constant argument, summarizing all the previous work concerning stability, oscillation properties, and existence of periodic solutions. In a series of papers [21, 22, 23, 24], Papaschinopoulos studied the topology equivalence, asymptotic behavior, and integral manifolds for these equations. The study of the existence of almost periodic solutions for EPCA has been studied (see [26, 27, 33, 34, 35]). Recently, quasi periodic solutions and pseudo almost periodic solutions have been studied (see [11, 36, 1]).

In this note, we consider the neutral differential equation with piecewise constant argument of the form

$$\dot{x}(t) = Ax(t) + \sum_{k=-N}^{N} A_k x([t+k]) + f(t), \quad x(t) \in \mathbb{R}^m, \tag{1.1}$$

where N is a given positive integer, A, A_k are given real $m \times m$-matrices, f is a bounded function on $\mathbb{R}$, $[\cdot]$ is the largest integer function.

The main technique of this note is to use the notion of spectrum of a function which has been wideespreadly employed in recent researches. We will estimate the spectrum of a bounded function (Carleman, Bohr spectrum), then we will consider the almost periodicity, periodicity, quasi periodicity of solutions. The main results of note are Theorem 3.6, Corollary 3.8, Corollary 3.9, Theorem 3.10 and Theorem 3.12. The estimates of the spectrum of a bounded solution which are obtained in Lemma 3.4, Lemma 3.11 are important. Theorem 3.6, Corollary 3.8 give the spectral conditions for almost periodicity of bounded solutions to Eq. (1.1). Corollary 3.9 gives a spectral condition for periodicity of bounded solutions to Eq. (1.1). Theorem 3.10 shows the existence of periodic solutions of Eq. (1.1). Theorem 3.12 gives a spectral condition for quasi periodicity of bounded solutions to Eq. (1.1).

2. Preliminaries

2.1. Notation

Thoughout the note $\mathbb{Z}$, $\mathbb{R}$, $\mathbb{C}$ stand for the sets of integer, real, complex numbers, respectively. $L^\infty(\mathbb{R}, \mathbb{R}^m)$, $BC(\mathbb{R}, \mathbb{R}^m)$, $BUC(\mathbb{R}, \mathbb{R}^m)$, $AP(\mathbb{R}^m)$ denote the spaces of all $\mathbb{R}^m$-valued essential bounded, bounded continuous, bounded uniformly continuous functions on $\mathbb{R}$ and their subspace including continuous and almost periodic functions, respectively. If A is a $m \times m$-matrix, then the notations $\sigma(A)$, $\rho(A)$, and $R(\lambda, A)$ stand for the spectrum, resolvent set, and resolvent of the matrix A. $sp_C(f)$, $\sigma_b(f)$ denote the Carleman, Bohr spectrum of f, respectively.

2.2. Spectral Theory of Functions

In this note, we will use the notions of Carleman spectrum and Bohr spectrum of a function.

For an essential bounded function $u : \mathbb{R} \to \mathbb{R}^m$, we define Fourier-Carleman transform of u

$$\widehat{u}(\lambda) = \begin{cases} \int\limits_0^\infty e^{-\lambda t} u(t)dt, & Re(\lambda) > 0 \\ -\int\limits_0^\infty e^{\lambda t} u(-t)dt, & Re(\lambda) < 0 \end{cases}.$$

Obvious, $\widehat{u}(\lambda)$ is holomorphic (analytic) in $\mathbb{C} \setminus i\mathbb{R}$. Thus, one get the concept of the Carleman spectrum of u as follows

Definition 2.1. We denote $sp_C(u) = \{\xi \in \mathbb{R} : \widehat{u}(\lambda)$ is not holomorphic in a some neighbourhood of $i\xi\}$, which is said to be Carleman spectrum of u.

Proposition 2.2. Let $\alpha, \beta \in \mathbb{R}$, $u, v \in L^\infty(\mathbb{R}, \mathbb{R}^m)$. Then

(i) (Linearity) $\widehat{(\alpha u + \beta v)}(\lambda) = \alpha\widehat{u}(\lambda) + \beta\widehat{v}(\lambda)$.

(ii) (Transform of derivation) Assume there exists $\dot{u} \in L^\infty(\mathbb{R}, \mathbb{R}^m)$. Then

$$\widehat{\dot{u}}(\lambda) = \lambda\widehat{u}(\lambda) - u(0).$$

(iii) Assume A is a $m \times m-$matrix, $u \in L^\infty(\mathbb{R}, \mathbb{R}^m)$. Then

$$\widehat{Au}(\lambda) = A\widehat{u}(\lambda).$$

Proposition 2.3 (see [25], page 20). Assume $u, v \in L^\infty(\mathbb{R}, \mathbb{R}^m)$, $\alpha \in \mathbb{C} \setminus \{0\}$, $\tau \in \mathbb{R}$. Then

(i) $sp_C(u)$ is close.

(ii) $sp_C(S(\tau)u) = sp_C(u)$.

(iii) $sp_C(\alpha u) = sp_C(u)$.

(iv) $sp_C(u + v) \subset sp_C(u) \cup sp_C(v)$.

(v) $sp_C(\dot{u}) \subset sp_C(u)$ if $\dot{u} \in L^\infty(\mathbb{R}, \mathbb{R}^m)$.

(vi) If A is a $m \times m$-matrix then $sp_C(Au) \subset sp_C(u)$.

For almost periodic functions (defined later), besides Carleman spectrum one has the concept of Bohr spectrum. Denote by $a(f, \lambda)$ the Bohr transformation of f, that is

$$a(f, \lambda) = \lim_{T \to +\infty} \frac{1}{2T} \int\limits_{-T}^{T} e^{-i\lambda t} f(t)dt.$$

We define the Bohr spectrum of f as follows

Definition 2.4. Give a function $f \in AP(\mathbb{R}^m)$. Let $\lambda \in \mathbb{R}$, then the mean value $a(f, \lambda) := \lim\limits_{T \to \infty} \frac{1}{2T} \int\limits_{-T}^{T} e^{-i\lambda t} f(t)dt$ exists and we call λ Fourier exponents of f, $a(f, \lambda)$ Fourier coefficients of f corresponding to λ. We denote $\sigma_b(f) = \{\lambda \in \mathbb{R} : a(f, \lambda) \neq 0\}$ and is said to be Bohr spectrum of f.

2.3. Almost Periodic Functions and Sequences

In the following, we give some concepts of almost periodic functions and sequences.

Definition 2.5. A subset $E \subset \mathbb{R}$ is called a relatively dense set in $\mathbb{R}$ if there exists a number $l > 0$ (inclusion length) such that $[a, a+l] \cap E \neq \emptyset \quad \forall a \in \mathbb{R}$.

Definition 2.6. A subset $E \subset \mathbb{Z}$ is called a relatively dense set in $\mathbb{Z}$ if there exists a number $l > 0$ (inclusion length) such that $[a, a+l] \cap E \neq \emptyset \quad \forall a \in \mathbb{Z}$.

Definition 2.7. Give a function $f : \mathbb{R} \to \mathbb{R}^m$ and $\varepsilon > 0$. We call the set $T(f, \varepsilon) = \{\tau \in \mathbb{R} : \|f(t+\tau) - f(t)\| < \varepsilon, \quad \forall t \in \mathbb{R}\}$ the ε-translation set of f and τ is called an $\varepsilon-$period for f.

Definition 2.8. Give a sequence $f : \mathbb{Z} \to \mathbb{R}^m$ and $\varepsilon > 0$. We call the set $TD(f, \varepsilon) = \{\tau \in \mathbb{Z} : \|f(t+\tau) - f(t)\| < \varepsilon, \quad \forall t \in \mathbb{Z}\}$ the ε-translation set of f and τ is called an $\varepsilon-$period for f.

Definition 2.9. A function $f : \mathbb{R} \to \mathbb{R}^m$ is called an almost periodic function if for all $\varepsilon > 0$ the set $T(f, \varepsilon)$ is a relatively dense set in $\mathbb{R}$.

Definition 2.10. A sequence $f : \mathbb{Z} \to \mathbb{R}^m$ is called an almost periodic sequence if for all $\varepsilon > 0$ the $TD(f, \varepsilon)$ is a relatively dense set in $\mathbb{Z}$.

Lemma 2.11. *If $f(t)$ is a continuous almost periodic function, then $\forall \varepsilon > 0, T(f, \varepsilon) \cap \mathbb{Z}$ is a relatively dense set in $\mathbb{Z}$.*

Proof. Using the Bochner criterion for a continuous, almost periodic function $f(t)$ we have the set $\{f_\tau(\cdot), \tau \in \mathbb{R}\}$ is relatively compact in $BC(\mathbb{R}, \mathbb{R}^m)$. It follows that the set $K = \{f_\tau(\cdot), \tau \in \mathbb{Z}\}$ is relatively compact in $BC(\mathbb{R}, \mathbb{R}^m)$. Thus, for all $\varepsilon > 0$, there exists $\{\tau_1, \ldots, \tau_n\} \subset \mathbb{Z}$ such that $\{f_{\tau_1}, \ldots, f_{\tau_n}\}$ is an $\varepsilon-$net of K.
Putting $l := \max\limits_{k:=1\ldots n} \{|\tau_k|\}$, one will prove the inclusion length for ε is $2l$.
In fact, $\forall t \in \mathbb{Z}, \exists \tau_k : \|f_t - f_{\tau_k}\| < \varepsilon$. Setting $\tau := t - \tau_k$ one has

i) $\tau \in \mathbb{Z}$,

ii) $\tau \in [t-l, t+l]$,

iii) $\|f_\tau - f\| = \|f_t - f_{\tau_k}\| < \varepsilon$,

i.e. $\tau \in (T(f, \varepsilon) \cap \mathbb{Z}) \cap [t-l, t+l]$.

So $T(f, \varepsilon) \cap \mathbb{Z}$ is relatively dense in $\mathbb{Z}$. □

Corollary 2.12. If $f(t)$ is a continuous almost periodic function, then $\{f(n), n \in \mathbb{Z}\}$ is an almost periodic sequence.

Proof. This has proved by remark that $T(f, \varepsilon) \cap \mathbb{Z} \subset TD(f, \varepsilon)$. □

Corollary 2.13. If $f(t)$ is a continuous almost periodic function, then $\bar{f}(t) = f([t])$ is an almost periodic function.

Proof. Corollary has followed by

$$\forall \tau \in T(f, \varepsilon) \cap \mathbb{Z} \Rightarrow \|\bar{f}(t+\tau) - \bar{f}(t)\| = \|f([t]+\tau) - f([t])\| \leq \varepsilon, \quad \forall t \in \mathbb{R}.$$

□

Lemma 2.14. *If $f(t)$ is a continuous almost periodic function, then $\{h_n\}$ is an almost periodic sequence and $h(t)$ is an almost periodic function.*

where $h(t) = -\int_t^{t+1} e^{A(t+1-s)} f(s)ds, \quad t \in \mathbb{R}$,

and $h_n = h(n), \quad n \in \mathbb{Z}$.

Proof. For $\tau \in T(\varepsilon, f)$ one has

$$\begin{aligned} \|h(t+\tau) - h(t)\| &= \| \int_{t+\tau}^{t+\tau+1} e^{A(t+1-s)} f(s)ds - \int_t^{t+1} e^{A(t+1-s)} f(s)ds \| \\ &= \| \int_0^1 e^{A(1-s)} [f(t+s+\tau) - f(t+s)] ds \| \\ &\leq \int_0^1 \|e^{A(1-s)}\| \|f(t+s+\tau) - f(t+s)\| ds \\ &\leq M\varepsilon, \end{aligned}$$

where $M := e^{\|A\|}$.
Because $T(\varepsilon, f)$ is relatively dense in $\mathbb{R}$, $h(t)$ is an almost periodic function. Thus, from Corollary 2.12 above $\{h_n\}$ is an almost periodic sequence and one has the proof. □

We have some conditions for the almost periodicity, periodicity of a bounded uniformly continuous function as follows.

Proposition 2.15 (see [2], page 322). Give a function $f \in BUC(\mathbb{R}, \mathbb{R}^m)$. If $sp_C(f)$ is discrete then f is almost periodic.

Lemma 2.16. *Assume $f \in BUC(\mathbb{R}, \mathbb{R}^m)$ and $sp_C(f)$ is countable. Then, f is almost periodic.*

Lemma 2.17 (see [2], page 323). *Let $f \in L^\infty(\mathbb{R}, \mathbb{R}^m)$ and $\tau > 0$. Then, f is τ−periodic if and only if $sp_C(f) \subset 2\pi\mathbb{Z}/\tau$.*

3. Main Results

In this section, we will deal with almost periodicity, periodicity, and quasi periodicity of bounded solutions and the existence of periodic solutions to Eq. (1.1) which is a so-called Massera criterion. First, we take a precise definition of solutions of Eq. (1.1).

Definition 3.1. A function $x : \mathbb{R} \to \mathbb{R}^m$ is said to be a solution of Eq. (1.1) if

(i) x is continuous on $\mathbb{R}$.

(ii) The derivative $\dot{x}(t)$ of $x(t)$ exists everywhere, with the possible exception of the points $n \quad (n \in \mathbb{Z})$, where one-sided derivatives exist.

(iii) $x(t)$ satisfies Eq. (1.1) on each interval $(n, n+1), \forall n \in \mathbb{Z}$.

3.1. The Characteristic Equation

Assume x is a bounded solution of Eq. (1.1). By using a variation of constants formula Eq. (1.1) is equivalent with

$$\begin{aligned} x(t) =& e^{A(t-n)}x(n) + \sum_{k=-N}^{N} \int_n^t e^{A(t-s)} ds A_k x(n+k) \\ &+ \int_n^t e^{A(t-s)} f(s) ds, \quad (n \le t < n+1; \forall n \in \mathbb{Z}). \end{aligned} \tag{3.2}$$

Because $x(t)$ is continuous on $\mathbb{R}$, by limiting when $t \to n+1$ one obtains the corresponding difference equation

$$\begin{aligned} x(n+1) =& e^{A}x(n) + \sum_{k=-N}^{N} \int_0^1 e^{A(1-s)} ds A_k x(n+k) \\ &+ \int_n^{n+1} e^{A(n+1-s)} f(s) ds, \quad (\forall n \in \mathbb{Z}). \end{aligned} \tag{3.3}$$

Setting

$$\begin{aligned} B_0 &= e^A + \int_0^1 e^{A(1-s)} ds A_0, \\ B_1 &= \int_0^1 e^{A(1-s)} ds A_1 - I, \\ B_k &= \int_0^1 e^{A(1-s)} ds A_k, \quad k \in [-1, \pm 2, \ldots, \pm N], \\ h_n &= -\int_n^{n+1} e^{A(n+1-s)} f(s) ds, \quad \forall n \in \mathbb{Z}. \end{aligned}$$

Eq. (3.3) become

$$\sum_{k=-N}^{N} B_k x(n+k) = h_n, \quad \forall n \in \mathbb{Z}. \tag{3.4}$$

The corresponding homogeneous equation of Eq.(3.4) is

$$\sum_{k=-N}^{N} B_k x(n+k) = 0, \quad \forall n \in \mathbb{Z}. \tag{3.5}$$

We will seek solutions of Eq. (3.5) as $x(n) = e^{n\lambda} c_\lambda$, where $\lambda \in \mathbb{C}$ satisfies

$$\det \left[\sum_{k=-N}^{N} B_k e^{(n+k)\lambda} \right] = 0, \quad \forall n \in \mathbb{Z}, \tag{3.6}$$

and c_λ is a m-dimention constant vector such that

$$\sum_{k=-N}^{N} B_k e^{(n+k)\lambda} c_\lambda = 0, \quad \forall n \in \mathbb{Z}. \tag{3.7}$$

Therefore, the characteristic equation associated with Eq. (1.1) is

$$\det\left[\sum_{k=-N}^{N} B_k e^{k\lambda}\right] = 0. \tag{3.8}$$

In order to comfortable for estimating the spectrum in the following we can give out some notions which relate rigorous to solutions of characteristic equation above. We call the set of solutions of this characteristic equation the spectrum of Eq. (1.1), denoted by

$$\Delta = \left\{\lambda \in \mathbb{C} : \det\left[\sum_{k=-N}^{N} B_k e^{k\lambda}\right] = 0\right\}.$$

Note that $\lambda \in \Delta$ if and only if e^{λ} is a solution of a polynomial of which order is not larger $2Nm$. Thus, e^{λ} has not larger $2Nm$ values and thus Δ is a discrete set of form $\{\lambda_k + 2\pi i\mathbb{Z}, \lambda_k \in \mathbb{C}, k = \overline{1, 2Nm}\}$.

The image spectrum of equation is a relative interesting character in estimates the spectrum and is defined by

$$\Delta_i = \{\xi \in \mathbb{R} : i\xi \in \Delta\}.$$

Using the spectrum to decomposition of functions is a method which is interested by many people (see [2, 12, 15, 18, 29]). The criterions for periodicity, almost periodic, quasi periodicity, pseudo almost periodicity ..., of bounded solutions is considered the improve of Massera theorem for general equations and is extented to the theory of functional spaces of admissibility. For EPCA of form (1.1), one obtains also the criterions for almost periodicity, periodicity, quasi periodicity of bounded solutions is due the method of spectral (see Theorem 3.6, Corollary 3.8, Corollary 3.9, Theorem 3.10 and Theorem 3.12).

Remark 3.2. In Fourier-Carleman transform, we only consider $\{\lambda : Re\lambda > 0\}$, otherwise $\{\lambda : Re\lambda < 0\}$ is presented similarly.

An important property of Eq. (1.1) is: each bounded solution is uniformly continuous. This is presented in the following lemma.

Lemma 3.3.

Assume x is a bounded solution of Eq. (1.1). Then $x \in BUC(\mathbb{R}, \mathbb{R}^m)$.

Proof. Put

$$\begin{aligned} M_1 &:= max(\|A\|, \|A_{-N}\|, \ldots, \|A_N\|), \\ M_2 &:= \sup_{t\in\mathbb{R}} \|x(t)\|, \\ M_3 &:= \sup_{t\in\mathbb{R}} \|f(t)\|, \\ M &:= M_1M_2 + (2N+1)M_1M_2 + M_3. \end{aligned}$$

Uniformly continuity on $\mathbb{R}$ of a bounded solution $x(t)$ follows from

$$\begin{aligned}\|x(t_2)-x(t_1)\| &= \|\int_{t_1}^{t_2}\dot{x}(t)dt\| \\ &\le \int_{t_1}^{t_2}\|\dot{x}(t)\|dt \\ &= \int_{t_1}^{t_2}\|Ax(t)+\sum_{k=-N}^{N}A_k x([t+k])+f(t)\|dt \\ &\le (t_2-t_1)M,\end{aligned}$$

for every $t_1 < t_2$ on $\mathbb{R}$. Lemma has been proved. □

Assume $x(t)$ is a bounded solution of Eq. (1.1), we follow $x \in BUC(\mathbb{R}, \mathbb{R}^m)$. Therefore we obtain an expression:

$$\lambda\widehat{x}(\lambda)-x(0) = A\widehat{x}(\lambda)+\sum_{k=-N}^{N}A_k g_k(\lambda)+\widehat{f}(\lambda), \tag{3.9}$$

where

$$\begin{aligned}g_k(\lambda) &= \int_0^\infty e^{-\lambda t}x([t+k])dt \\ &= \frac{1-e^{-\lambda}}{\lambda}\sum_{n=0}^{\infty}e^{-n\lambda}x(n+k).\end{aligned} \tag{3.10}$$

On the other hand, from Eq. (3.3) it follows

$$\begin{aligned}x([t+1]) =& e^A x([t])+\sum_{k=-N}^{N}\int_0^1 e^{A(1-s)}dsA_k x([t+k]) \\ &+\int_{[t]}^{[t+1]}e^{A([t+1]-s)}f(s)ds.\end{aligned} \tag{3.11}$$

By Fourier-Carleman transform, we have that

$$g_1(\lambda) = e^A g_0(\lambda)+\sum_{k=-N}^{N}\int_0^1 e^{A(1-s)}dsA_k g_k(\lambda)-\widehat{\bar{h}}(\lambda), \tag{3.12}$$

which is equivalent with

$$\sum_{k=-N}^{N}B_k g_k(\lambda) = \widehat{\bar{h}}(\lambda), \tag{3.13}$$

where

$$\bar{h}(t) = h([t]) = -\int_{[t]}^{[t+1]}e^{A([t+1]-s)}f(s)ds.$$

Simple transforms for $g_k(\lambda)$

$$
\begin{aligned}
g_k(\lambda) &= \frac{1-e^{-\lambda}}{\lambda}\sum_{n=0}^{\infty} e^{-n\lambda}x(n+k) \\
&= \begin{cases} \frac{1-e^{-\lambda}}{\lambda}e^{k\lambda}[\frac{\lambda}{1-e^{-\lambda}}g_0(\lambda) - x(0) - \cdots - e^{-(k-1)\lambda}x(k-1)], & k \geq 1 \\ \frac{1-e^{-\lambda}}{\lambda}e^{k\lambda}[\frac{\lambda}{1-e^{-\lambda}}g_0(\lambda) + e^{-k\lambda}x(k) + \cdots + e^{\lambda}x(-1)], & k \leq -1 \end{cases} \\
&= \begin{cases} e^{k\lambda}g_0(\lambda) - \frac{1-e^{-\lambda}}{\lambda}e^{k\lambda}[x(0) - \cdots - e^{-(k-1)\lambda}x(k-1)], & k \geq 1 \\ e^{k\lambda}g_0(\lambda) + \frac{1-e^{-\lambda}}{\lambda}e^{k\lambda}[e^{-k\lambda}x(k) + \cdots + e^{\lambda}x(-1)], & k \leq -1 \end{cases} \\
&= e^{k\lambda}g_0(\lambda) + p_k(\lambda), \quad \text{with } p_k(\lambda) \text{ is holomorphic in } \lambda \text{ on } \mathbb{C}.
\end{aligned}
$$

Replacing in Eq. (3.9), and Eq. (3.13) we obtain a system

$$
\begin{aligned}
\sum_{k=-N}^{N} A_k e^{k\lambda} g_0(\lambda) &= (\lambda - A)\widehat{x}(\lambda) - x(0) - \widehat{f}(\lambda) - \sum_{k=-N}^{N} A_k p_k(\lambda) && (3.14)\\
&= (\lambda - A)\widehat{x}(\lambda) - \widehat{f}(\lambda) + p(\lambda) && (3.15)\\
\sum_{k=-N}^{N} B_k e^{k\lambda} g_0(\lambda) &= \widehat{\bar{h}}(\lambda) - \sum_{k=-N}^{N} B_k p_k(\lambda) && (3.16)\\
&= \widehat{\bar{h}}(\lambda) + q(\lambda), && (3.17)
\end{aligned}
$$

where p, q are vector-value functions and holomorphic in λ on $\mathbb{C}$.

3.2. Criterion of Almost Periodicity of a Bounded Solution

From the expressions obtained above we have an estimate of the Carleman spectrum of a bounded solution.

Lemma 3.4. *Assume x is a bounded solution of Eq. (1.1). Then we have the following spectral estimate:*

$$
sp_C(x) \subset sp_C(f) \cup sp_C(\bar{h}) \cup \sigma_i(A) \cup \Delta_i, \tag{3.18}
$$

where $\sigma_i(A) = \{\xi \in \mathbb{R} : i\xi \in \sigma(A)\}$.

Proof. Assume a real number $\xi \notin sp_C(f) \cup sp_C(\bar{h}) \cup \sigma_i(A) \cup \Delta_i$, we will prove that $\xi \notin sp_C(x)$. In fact, from assumption it follows that $\widehat{f}(\lambda), \widehat{\bar{h}}(\lambda)$ is holomorphic in a some neighbourhood of $i\xi$ and resolvent operator $R(\lambda, A)$ is holomorphic in a some neighbourhood of $i\xi$. On the other hand, as $\xi \notin \Delta_i$ so $(\sum_{k=-N}^{N} B_k e^{k\lambda})^{-1}$ is holomorphic in a neighbourhood of $i\xi$. From Eq. (3.17) one has

$$
g_0(\lambda) = (\sum_{k=-N}^{N} B_k e^{k\lambda})^{-1}(\widehat{\bar{h}}(\lambda) + q(\lambda)),
$$

then replace in Eq. (3.15) one obtains

$$\sum_{k=-N}^{N} A_k e^{k\lambda}(\sum_{k=-N}^{N} B_k e^{k\lambda})^{-1}(\widehat{\bar{h}}(\lambda)+q(\lambda)) = (\lambda - A)\widehat{x}(\lambda) - \widehat{f}(\lambda) + p(\lambda),$$

is equivalent with

$$\widehat{x}(\lambda) = R(\lambda, A)(r(\lambda) + \widehat{f}(\lambda) + u(\lambda)\widehat{\bar{h}}(\lambda)),$$

where r, u are vector, matrix-valued functions (respectively) which are holomorphic in λ on $\mathbb{C}$ i.e. $\widehat{x}(\lambda)$ is holomorphic in a some neighbourhood of $i\xi$ i.e. $\xi \notin sp_C(x)$.
Lemma has been proved. □

We have some first remarks of the right side of Eq. (3.18):

Remark 3.5. i) $\sigma_i(A) = \{\xi \in \mathbb{R} : i\xi \in \sigma(A)\}$ is a finite set on $\mathbb{R}$.

ii) $\Delta_i = \{\xi \in \mathbb{R} : i\xi \in \Delta\}$ is a discrete set of form

$$\{\xi_k + 2\pi\mathbb{Z}, \xi_k \in \mathbb{R}, k \in \{1, \ldots, 2Nm\}\}.$$

From lemmas above one gives criterions of almost periodicity of a bounded solution as follows.

Theorem 3.6. *If a function f satisfies that $sp_C(f), sp_C(\bar{h})$ are countable then each bounded solution of Eq. (1.1) is an almost periodic solution.*

Proof. Assume x is a bounded solution of Eq. (1.1). From lemmas 3.3, 3.4 and Remark 3.5 it follows that $x \in BUC(\mathbb{R}, \mathbb{R}^m)$ and $sp_C(x)$ is countable. Lemma 2.16 give that $x(t)$ will an almost periodic solution. □

We have a relation between $sp_C(f)$ and $sp_C(\bar{h})$ as follows

Lemma 3.7. *If $sp_C(f)$ is discrete and the representation $f(t) = \sum_{n\geq 1} a_n e^{i\lambda_n t}$ for $\sum_{n\geq 1} \|a_n\| < \infty, a_n \neq 0$ then $sp_C(\bar{h})$ is also discrete. Moreover, one has an estimate*

$$sp_C(\bar{h}) \subset \{\lambda_n + 2\pi\mathbb{Z}, n > 1\}.$$

Proof. We have $f(t) = \sum_{n\geq 1} a_n e^{i\lambda_n t}$ for $\sum_{n\geq 1} \|a_n\| < \infty, a_n \neq 0$.

Then,

$$\begin{aligned}\bar{h}(t) &= \int_0^1 e^{A(1-s)} f([t]+s)ds \\ &= \sum_{n\geq 1} \int_0^1 e^{A(1-s)} e^{i\lambda_n s} ds a_n e^{i\lambda_n [t]} \\ &= \sum_{n\geq 1} b_n e^{i\lambda_n [t]},\end{aligned}$$

where $b_n = \int_0^1 e^{A(1-s)} e^{i\lambda_n s} ds a_n$. Thus, for $Re\lambda > 0$

$$\begin{aligned}\int_0^\infty e^{-\lambda t}\bar{h}(t)dt &= \sum_{n\geq 1} b_n \int_0^\infty e^{-\lambda t + i\lambda_n [t]} dt \\ &= \sum_{n\geq 1} b_n \sum_{m\geq 0} \int_m^{m+1} e^{-\lambda t + i\lambda_n m} dt \\ &= \sum_{n\geq 1} b_n \sum_{m\geq 0} e^{(i\lambda_n - \lambda)m} \frac{1-e^{-\lambda}}{\lambda} \\ &= \sum_{n\geq 1} b_n \frac{1}{1-e^{(i\lambda_n-\lambda)}} \frac{1-e^{-\lambda}}{\lambda}.\end{aligned}$$

For $b_n \neq 0$, it is easy to see $\frac{1}{1-e^{(i\lambda_n-\lambda)}}\frac{1-e^{-\lambda}}{\lambda}$ is holomorphic in a small enough neighbourhood of $i\xi$ for $\xi \notin \{\lambda_n + 2\pi\mathbb{Z}, n \geq 1\}$ (because $\{\lambda_1, \lambda_2, \ldots\}$ is discrete) i.e. $sp_C(\bar{h}) \subset \{\lambda_n + 2\pi\mathbb{Z}, n \geq 1\}$.

Thus, $sp_C(\bar{h})$ is discrete, one has the proof. □

Therefore, we have a better estimate for almost periodicity of a bounded solution.

Corollary 3.8. If f satisfies that $sp_C(f)$ is discrete and the representation $f(t) = \sum_{n\geq 1} a_n e^{i\lambda_n t}$ for $\sum_{n\geq 1} \|a_n\| < \infty, a_n \neq 0$ then a bounded solution of Eq. (1.1) will be an almost periodic solution.

3.3. Criterion of Periodicity of a Bounded Solution

From estimate the spectrum above, by the condition "the characteristic equation (3.8) has no solutions on the imaginary axis" and "$\sigma(A) \cap i\mathbb{R} \subset 2\pi\mathbb{Z}$" we have a corollary of periodic solutions.

Corollary 3.9. By the condition "the characteristic equation (3.8) has no solutions on the imaginary axis" and "$\sigma(A) \cap i\mathbb{R} \subset 2\pi\mathbb{Z}$" we have
1) If f is periodic of period $p \in \mathbb{N}$ then a bounded solution x is also periodic with the same period (called harmonic solution).
2) If f is periodic of period p/q (where $p, q \in \mathbb{N}$ are relatively prime) then a bounded solution x is periodic of period p (called subharmonic solution).

Proof. Function $\bar{h}$ is periodic of period p because

$$\begin{aligned}\bar{h}(t+p) &= \int_0^1 e^{A(1-s)} f([t+p]+s)ds \\ &= \int_0^1 e^{A(1-s)} f([t]+p+s)ds \\ &= \int_0^1 e^{A(1-s)} f([t]+s)ds \text{ (as f is periodic of period } p, (p/q)) \\ &= \bar{h}(t).\end{aligned}$$

From lemma 2.17 it follows that $sp_C(f), sp_C(\bar{h}) \subset \frac{2\pi\mathbb{Z}}{p}$. Then, by the above conditions we have $\sigma_i(A) \cap \mathbb{R} \subset 2\pi\mathbb{Z}, \Delta_i = \emptyset$ and from Lemma 3.4 we have $sp_C(x) \subset \frac{2\pi\mathbb{Z}}{p}$ and thus x is periodic of period p. Corollary has been proved. □

3.4. Massera Criterion for the Existence of a Periodic Solution

Corollary 3.9 gives us a criterion (is called the Massera criterion) for the existence of periodic solutions as follows

Theorem 3.10. *Assume Eq. (1.1) satisfies conditions of Corollary 3.9 above. Assumption f is a periodic function with rational period $\tau = p/q$. Then Eq. (1.1) has a periodic solution with period p if and only if it has a bounded solution on the positive half line.*

Proof. Necessaty is obvious. We only have to prove sufficiency. However, from Corollary 3.9 any bounded solution is periodic of period p so we only have to prove: "if there exists a bounded solution on the positive half line then there exists also a bounded solution on the whole line $\mathbb{R}$". In fact, assume that x is a bounded solution on the positive half line of Eq. (1.1). From the proof of Lemma 3.3 we follow x is uniformly continuous on the positive half line. Put $x_n(t) = x(t+np)$, for $t \geq -np$. Using the Arzel-Ascoli's theorem one may assume that $y \in BUC(\mathbb{R}, \mathbb{R}^m)$ is a local uniformly limit of a sequence of functions $\{x_n\}$ on $\mathbb{R}$. Give $t \geq \tau \in \mathbb{R}$. For every n such that $np + \tau \geq 0$ one gets that

$$\begin{aligned}x_n(t) - x_n(\tau) &= x(np+t) - x(np+\tau) \\ &= \int_{np+\tau}^{np+t} (Ax(s) + \sum_{k=-N}^{N} A_k x([s+k]) + f(s))ds \\ &= \int_\tau^t (Ax(np+s) + \sum_{k=-N}^{N} A_k x([np+s+k]) + f(np+s))ds \\ &= \int_\tau^t (Ax_n(s) + \sum_{k=-N}^{N} A_k x_n([s+k]) + f(s))ds.\end{aligned}$$

Let $n \to \infty$, the above equation becomes

$$y(t) - y(\tau) = \int_{\tau}^{t} (Ay(s) + \sum_{k=-N}^{N} A_k y([s+k]) + f(s))ds.$$

So, y is a bounded solution on $\mathbb{R}$ of Eq. (1.1). Therefore theorem has been proved. □

3.5. Criterion for Quasi Periodicity of a Bounded Solution

Result of Theorem 3.6 above: "a bounded solution is almost periodic" helps us to estimate of Bohr spectrum of a bounded solution.

Lemma 3.11. *If $sp_C(f), sp_C(\bar{h})$ are countable then we have estimate the Bohr spectrum of a bounded solution x of Eq. (1.1) as follows*

$$\sigma_b(x) \subset \sigma_b(f) \cup \sigma_b(\bar{h}) \cup \sigma_i(A) \cup \Delta_i. \tag{3.19}$$

Proof. From Eq. (3.4) we have

$$\sum_{k=-N}^{N} B_k \bar{x}(t+k) = \bar{h}(t). \tag{3.20}$$

By the Theorem 3.6, a bounded solution x is an almost periodic solution. Thus, from corollary 2.13 $\bar{x}$ is also an almost periodic function so Bohr transforms of two sides of Eq. (1.1) exist. Therefore,

$$\begin{aligned}\lim_{T\to\infty} \frac{1}{2T}\int_{-T}^{T} e^{-i\lambda t}\dot{x}(t)dt = &\lim_{T\to\infty} \frac{1}{2T}\int_{-T}^{T} e^{-i\lambda t}Ax(t)dt \\ &+ \lim_{T\to\infty} \frac{1}{2T}\int_{-T}^{T} e^{-i\lambda t}\sum_{k=-N}^{N} A_k\bar{x}(t+k)dt \\ &+ \lim_{T\to\infty} \frac{1}{2T}\int_{-T}^{T} e^{-i\lambda t}f(t)dt.\end{aligned} \tag{3.21}$$

Transform of the left side of Eq. (3.21) by the integrate by parts formula and note that x and $e^{-i\lambda t}$ are bounded we obtain:

$$\begin{aligned}\lim_{T\to\infty} \frac{1}{2T}\int_{-T}^{T} e^{-i\lambda t}\dot{x}(t)dt = &\lim_{T\to\infty} \frac{1}{2T} e^{-i\lambda t}x(t)|_{-T}^{T} \\ &-(-i\lambda)\lim_{T\to\infty} \frac{1}{2T}\int_{-T}^{T} e^{-i\lambda t}x(t)dt \\ =&i\lambda \lim_{T\to\infty} \frac{1}{2T}\int_{-T}^{T} e^{-i\lambda t}x(t)dt \\ =&i\lambda a(x,\lambda).\end{aligned}$$

Similarly transform of the right side we obtain

$$\lim_{T\to\infty}\frac{1}{2T}\int_{-T}^{T} e^{-i\lambda t}Ax(t)dt = A\lim_{T\to\infty}\frac{1}{2T}\int_{-T}^{T} e^{-i\lambda t}x(t)dt = Aa(x,\lambda)$$

$$\begin{aligned}\lim_{T\to\infty}\frac{1}{2T}\int_{-T}^{T} e^{-i\lambda t}\sum_{k=-N}^{N}A_k\bar{x}(t+k)dt &= \sum_{k=-N}^{N}A_k\lim_{T\to\infty}\frac{1}{2T}\int_{-T}^{T} e^{-i\lambda t}\bar{x}(t+k)dt\\ &= \sum_{k=-N}^{N}A_k e^{ik\lambda}\lim_{T\to\infty}\frac{1}{2T}\int_{-T}^{T} e^{-i\lambda(t+k)}\bar{x}(t+k)dt\\ &= \sum_{k=-N}^{N}A_k e^{ik\lambda}\lim_{T\to\infty}\frac{1}{2T}\int_{-T+k}^{T+k} e^{-i\lambda t}\bar{x}(t)dt\\ &= \sum_{k=-N}^{N}A_k e^{ik\lambda}a(\bar{x},\lambda)\end{aligned}$$

$$\lim_{T\to\infty}\frac{1}{2T}\int_{-T}^{T} e^{-i\lambda t}f(t)dt = a(f,\lambda).$$

So, Eq. (3.21) will become

$$(i\lambda - A)a(x,\lambda) = \sum_{k=-N}^{N}A_k e^{ik\lambda}a(\bar{x},\lambda) + a(f,\lambda). \tag{3.22}$$

On the other hand, from lemma 2.14 function $h(t)$ is almost periodic and from Corollary 2.13 $\bar{h}(t)$ is almost periodic so Bohr transforms of two sides of Eq. (3.20) exist. Thus, by similarly transform we obtain

$$\sum_{k=-N}^{N}B_k e^{ik\lambda}a(\bar{x},\lambda) = a(\bar{h},\lambda). \tag{3.23}$$

From Eq. (3.23) and the condition $\lambda \notin \Delta_i$, i.e. $\sum_{k=-N}^{N}B_k e^{ik\lambda}$ is invertible, we follow $a(\bar{x},\lambda) = 0$ provided that $a(\bar{h},\lambda) = 0$.

For a real number $\lambda \notin \sigma_b(f)\cup\sigma_b(\bar{h})\cup\sigma_i(A)\cup\Delta_i$ we have $a(f,\lambda) = 0, a(\bar{h},\lambda) = 0$ and $(i\lambda - A)$ is invertible thus from Eq. (3.22) it follows that $a(x,\lambda) = 0$ or $\lambda \notin \sigma_b(x)$. So

$$\sigma_b(x) \subset \sigma_b(f)\cup\sigma_b(\bar{h})\cup\sigma_i(A)\cup\Delta_i$$

and lemma has been proved. □

From Lemma 3.11 and Remark 3.5 we obtain a criterion for quasi periodicity of a bounded solution.

Theorem 3.12. *If $f, \bar{h}$ are quasi periodic functions then each bounded solution of Eq. (1.1) is also quasi periodic.*

Proof. From the assumption of quasi periodicity of $f, \bar{h}$, it follows that $\sigma_b(f), \sigma_b(\bar{h})$ have bases which are integer and finite. Assume that x is a bounded solution of Eq. (1.1). By estimating of spectrum in Lemma 3.11 and Remark 3.5 it follows that $\sigma_b(x)$ has also a basis which is integer and finite. Therefore, the bounded solution x is quasi periodic. Theorem has been proved. □

To illustrate the interesting property of the above Massera criterion for the periodicity of solutions of differential equations with piecewise constant argument we will consider the following examples

Example 3.13.

Consider the following equations

$$\text{(i)}\ \dot{x}(t) = -x([t]) + \begin{bmatrix} cos(2\pi t) \\ cos(2\pi t) \end{bmatrix}.$$

$$\text{(ii)}\ \dot{x}(t) = -x([t]) + \begin{bmatrix} cos(4\pi t) \\ cos(4\pi t) \end{bmatrix}.$$

$$\text{(iii)}\ \dot{x}(t) = -x([t]) + \begin{bmatrix} cos(t) \\ cos(t) \end{bmatrix}.$$

In this case
The characteristic equation (3.8) now is: $det[e^{\lambda}] = 0$. It has no solutions, so it has no solutions on the imaginary axis. Simultaneously, $\sigma(A) \cap i\mathbb{R} = \{0\} \subset 2\pi\mathbb{Z}$.

Therefore the conditions of Corollary 3.9 are satisfied. It follows that

Eq.(i) gives us a unique harmonic solution, that is

$$x(t) = \frac{1}{2\pi} \begin{bmatrix} sin(2\pi t) \\ sin(2\pi t) \end{bmatrix}.$$

Eq.(ii) gives us a unique subharmonic solution, that is

$$x(t) = \frac{1}{4\pi} \begin{bmatrix} sin(4\pi t) \\ sin(4\pi t) \end{bmatrix}.$$

However, Eq.(iii) has a unique bounded solution (quasi periodic solution) which isn't periodic

$$x(t) = ([t] + 1 - t) \begin{bmatrix} sin([t]) - sin([t-1]) \\ sin([t]) - sin([t-1]) \end{bmatrix} + \begin{bmatrix} sin(t) - sin([t]) \\ sin(t) - sin([t]) \end{bmatrix}.$$

In fact, by analytic caculation we have $sp_C(x) = \{1+2\pi\mathbb{Z}\} \cup \{-1+2\pi\mathbb{Z}\}$. Assume that x is periodic with a some period $\tau > 0$. Then the Lemma 2.17 follows $sp_C(x) \subset \frac{2\pi\mathbb{Z}}{\tau}$. Thus we have $1 = \frac{2\pi n}{\tau}$ and $1 + 2\pi = \frac{2\pi k}{\tau}$ where $n, k \in \mathbb{N}$. This is a contradiction by now

$$2\pi = \frac{2\pi k}{\tau} - 1 = \frac{2\pi k}{2\pi n} - 1 = \frac{k}{n} - 1.$$

However, we have $\overline{\sigma_b(x)} = sp_C(x) = \{1 + 2\pi\mathbb{Z}\} \cup \{-1 + 2\pi\mathbb{Z}\}$, it follows that $\sigma_b(x) = \{1 + 2\pi\mathbb{Z}\} \cup \{-1 + 2\pi\mathbb{Z}\}$ i.e. the set of Fourier exponents of x has a basis $\{1, 1 + 2\pi\}$ which is integer and finite so x is a quasi periodic solution.

Therefore, the Massera theorem is true for f which is periodic of rational period but is not true for f which is periodic of irrational period.

Acknowledgments

The author is grateful to Professor Nguyen Van Minh for pointing out the problem and his constant interest in the work. He also thanks the referee for carefully reading the manuscript and for his valuable suggestions.

References

[1] A. I. Alonso, J. Hong, J. Rojo, A class of ergodic solutions of differential equation with piecewise continuous delays, *Dynam. Systems Appl.,* **7** (1998), 561-574.

[2] W. Arendt, C. J. K. Batty, M. Hieber, F. Neubrander, Vector-valued Laplace Transforms and Cauchy Problems, *in* "Monographs in Mathematics", Vol. **96**, Birkhäuser Verlag, Basel-Boston-Berlin 2001.

[3] S. Busenberg, K. L. Cooke, Models of vertically transmitted diseases with sequential-continuous dynamics, *in* "Nonlinear Phenonmena in Mathematical Sciences" (V. Lakshmikantham, Ed.), 179-187; Academic Press, New York, 1982.

[4] A. Cabada, J. B. Ferreiro, J. J. Nieto, Green's function and comparison principles for first order periodic differential equations with piecewise constant arguments, *J. Math. Anal. Appl.*, **291** (2004), 690-697.

[5] K. L. Cooke, J. Wiener, Retarded differential equations with piecewise constant delays, *J. Math. Anal. Appl.*, **99** 1984, 265-297.

[6] K. L. Cooke, J. Wiener, Neutral differential equations with piecewise constant argument, *Boll. Un. Mat. Ital.*, **7** 1987, 321-349.

[7] K. L. Cooke, J. Wiener, A survey of differential equation with piecewise continuous delays, in "Lecture Notes in Mathematics", Vol. **1475**, 1-15, Springer-Verlag, Berlin 1991.

[8] M. H. Eduardo, "*A Massera type criterion for a partial neutral functional differential equation*", Elec. J. Diff. Eq. Vol. **2002**, No. 40 (2002), 1-17.

[9] T. Furumochi, T. Naito, Nguyen Van Minh, Boundedness and almost periodicity of solutions of partial functional differential equations, *J. Differential Equations*, **180** (2002), 125-152.

[10] Y. Hino, T. Naito, Nguyen Van Minh, J. S. Shin, *Almost Periodic Solutions of Differential Equations in Banach Spaces*, Taylor and Francis Group, London-New York, 2002.

[11] Tassilo Küpper, Rong Yuan, On Quasi-Periodic Solutions of Differential Equations with Piecewise Constant Argument, *J. Math. Anal. Appl.*, **267** (2002), 173-193 .

[12] B. M. Levitan, V. V. Zhikov, "Almost periodic functions and differential equations", Monographs in Mathematics, Vol. **96**, (1978), Moscow Univ. Publ. House (English translation by Cambridge University Press, 1982).

[13] Gza Makay, *On some possible extension of Massera's theorem*, EJQTDE, Proc. 6th Coll. QTDE, 2000 No. **16**.

[14] J. L. Massera, The existence of periodic solutions of systems of differential equations, *Duke Math. J.*, **17** (1950), 457-475.

[15] N. M. Man, N. V. Minh, "*On the existence of quasi periodic and almost periodic solutions of neutral functional differential equations* ", (2003).

[16] S. Murakami, T. Naito, Nguyen Van Minh, Evolution semigroups and sums of commuting operators: A new approach to the admissibility theory of function spaces, *J. Differential Equations*, **164** (2000), 240-285.

[17] S. Murakami, T. Naito, Nguyen Van Minh, Massera Theorem for almost periodic solutions of functional differential equations, *J. Math. Soc. Japan*, **56** (2004), N.1, 247-268.

[18] T. Naito, Nguyen Van Minh, J. S. Shin, New spectral criteria for almost periodic solutions of evolution equations, *Studia Mathematica*, **145** (2001), 97-111.

[19] T. Naito, N. V. Minh, R. Miyazaki, Y. Hamaya, Boundedness and almost periodicity in dynamical systems, *J. Difference Equations Appls.*, **7** (2001), 507-527.

[20] T. Naito, R. Miyazaki, Nguyen Van Minh, J. S. Shin, A decomposition theorem for bounded solutions and the existence of periodic solutions to periodic equations, *J. Differential Equation*, **160** (2000), N.1, pp. 263-282.

[21] G. Papaschinopoulos, Some results concerning a class of differential equations with piecewise constant argument, *Math. Nachr.*, **166** (1994), 193-206.

[22] G. Papaschinopoulos, Exponential dichotomy, topological equivalence and structural stability for differential equations with piecewise constant argument, *Analysis*, **14** (1994), 239-247.

[23] G. Papaschinopoulos, On the integral manifold for a system of differential equations with piecewise constant argument, *J. Math. Anal. Appl.*, **201** (1996), 75-90.

[24] G. Papaschinopoulos, Linearization near the integral manifold for a system of differential equations with piecewise constant argument, *J. Math. Anal. Appl.*, **215** (1997), 317-333.

[25] J. Prüss, "Evolutionary integral equations and applications", Birkhauser, Basel (1993).

[26] George Seifert, Almost Periodic Solutions of Certain Differential Equations with Piecewise Constant Delays and Almost Periodic Time Dependence, *J. Differential Equation*, **164** (2000), 451-458 .

[27] George Seifert, Second order neutral delay differential equations with piecewise constant time dependence, *J. Math. Anal. Appl.*, (2003) **281** 1-9.

[28] S. M. Shah, J. Wiener, Advanced differential equations with piecewise constant argument deviations , *Int. J. Math. Math. Sci.*, **6** (1983), 671-703.

[29] Nguyen Trung Thanh, Massera criterion for periodic solutions of differential equations with piecewise constant argument, *J. Math. Anal. Appl.*, to appear.

[30] J. Wiener, "Generalized Solutions of Functional Differential Equations", World Scientific, Singapore, 1993.

[31] J. Wiener, K. L. Cooke, Oscillations in systems of differential equations with piecewise constant argument, *J. Math. Anal. Appl.*, **137** (1989), 221-239.

[32] J. Wiener, W. Heller, Oscillatory and periodic solutions to a diffusion equation of neutral type, *Internat.J.Math.&Math.Sci.*, Vol.**22**, No.2 (1999), 313-348.

[33] Rong Yuan, Jialin Hong, Almost periodic solutions of differential equations with piecewise constant argument, *Analysis*, **16** (1996), 171-180.

[34] Rong Yuan, Jialin Hong, The existence of almost periodic solution for a class of differential equations with piecewise constant argument, *Nonlinear Anal.*, **28** (1997), 1439-1450.

[35] Rong Yuan, Almost periodic solutions of a class of singularly perturbed differential equations with piecewise constant argument, *Nonlinear Anal.*, **37** (1999), 641-659.

[36] Rong Yuan, On the logistic delay differential equations with piecewise constant argument and quasi periodic time dependence, *J. Math. Anal. Appl.*, **274** (2002) 124-133.

[37] F. Zhang, A. Zhao, J. Yan, Monotone iterative method for differential equations with piecewise constant arguments, *Portugaliae Mathematica* Vol. **57** Fasc. 3-2000.

In: Focus on Evolution Equations
Editor: Gaston M. N'Guerekata

ISBN: 978-1-60021-342-7

Chapter 9

On the Asymptotic Equivalence of Solutions of the Linear Evolution Equations

Dang Dinh Chau[1,*] ***and Kieu Thu Linh***[2,†]
[1]Department of Mathematics, Hanoi University of Science,
334 Nguyen Trai, Hanoi, Vietnam
[2]Centre for Education Technology - CGD, 50 Lieu Giai,
Ba Dinh, Hanoi, Vietnam

Abstract

In this article we study conditions on the asymptotic equivalence of linear evolution equations in Banach spaces. Besides, we discuss the asymptotic relationship between C_0 semigroups and strongly continuous evolutionary process.

1. Introduction

We consider the following linear evolution equations

$$\frac{dx(t)}{dt} = Ax(t), t \geq 0, \tag{1.1}$$

and

$$\frac{dy(t)}{dt} = C(t)y(t), t \geq 0, \tag{1.2}$$

where $x(t), y(t) \in X$, X is a complex Banach space, A, $C(t)$ are linear operators acting on X for each $t \geq 0$. Under suitable conditions, Eq.(1.1) and Eq.(1.2) are well posed (see [1], [4], [7]).

*E-mail address: chaudd@vnu.edu.vn

†The authors are grateful to the referee for carefully reading the paper and suggestions to improve the presentation.

We recall that Eq.(1.1) and Eq.(1.2) are said to be asymptotically equivalent if there exists a bijection between the set of solutions of Eq.(1.1) and the one of Eq.(1.2) such that

$$\lim_{t\to\infty} \|x(t) - y(t)\| = 0. \tag{1.3}$$

An interesting problem in studying the qualitative theory of solutions to differential equations is to find conditions such that Eq.(1.1) and Eq.(1.2) are asymptotically equivalent. The first results of this problem were given by N. Levinson in 1946 (see [5]). Then, these results have been developed in many ways by many authors (see [2], [3], [6], [8], [10]). In this paper we study the asymptotic equivalence of linear evolution equations in Banach spaces. The obtained results are extension of Levinson's results to the case of general Banach spaces.

In case $X = H$ is a Hilbert space and $A \in L(H)$ has a triangle form (see [2], p.9), the sufficient conditions for Eq.(1.1) and Eq.(1.2) to be asymptotically equivalent can be replaced by weaker conditions. Moreover, the methods used in the proof of the results obtained in this part (part 3) can be applied to the study of the asymptotic behavior of solutions to differential equations in infinite dimensional spaces, thanks to this method, in many case, we can replace the study of the asymptotic behavior of solutions of infinite systems of differential equations by that of finite systems of differential equations .

2. The Asymptotic Equivalence of Strongly Continuous Evolutionary Processes

Throughout this paper, we consider the Eq.(1.1) and Eq.(1.2), where $C(t) = A + B(t)$, $A \in L(X)$,$B(.) : [0, +\infty) \to L(X)$ satisfy

$$\int_0^\infty \|B(\tau)\| d\tau < \infty. \tag{2.4}$$

Assume that $(T(t))_{t\in R}$ is a C_0 group generated by A. Correspondingly to Eq.(1.2), we usually consider the following integral equation

$$y(t) = T(t-s)y(s) + \int_s^t T(t-s)B(s)y(s)ds, t \geq s. \tag{2.5}$$

We can easily show that the solution of Eq.(2.2) can be written in the form

$$y(t) = Y(t,s)y_0 \quad (s \leq t \leq \infty), \quad x(s) = x_0,$$

where $(Y(t,s))_{t\geq s}$ is a family of bounded linear operators in X satisfying the equation

$$Y(t,s) = T(t-s) + \int_s^t T(t-\tau)B(\tau)Y(\tau,s)d\tau, t \geq s. \tag{2.6}$$

A two-parameter family $(U(t,s))_{t\geq s}$ of bounded linear operators on X is called a strongly continuous evolutionary process if it has the following four properties:

1. $U(t,t) = I, \quad$ for all $t \geq 0$

2. $U(t,s)U(s,\tau) = U(t,\tau)$, for all $t \geq s \geq \tau \geq 0$

3. The map $(t,s) \mapsto U(t,s)x$ is continuous for every fix $x \in X$

4. $\|U(t,s)\| < Ne^{\omega(t-s)}$, for some positive constants N, ω independent of $t \geq s \geq 0$

Definition 2.1. *$(T(t))_{t\geq 0}$ and $(U(t,s))_{t\geq s}$ are said to be asymptotically equivalent if for every $x_0 \in X$, there exists $y_0 \in X$, such that*

$$\lim_{t\to\infty} \|T(t-s)x_0 - U(t,s)y_0\| = 0 \tag{2.7}$$

for each fixed $s \in R$ and conversely.

To prove the first theorem, we need the following lemmas

Lemma 2.2. *If $(T(t))_{t\geq 0}$ is a uniformly bounded C_0 semigroup generated by A, then*
1. *The Cauchy problem related to Eq. (2.2) has a unique solution*
2. *There exists a constant $K \geq 0$ such that*

$$\|Y(t,s)\| < K, \forall t \geq s. \tag{2.8}$$

Proof. For the proof the reader can see [1], [4]. □

Lemma 2.3. *Let $(T(t))_{t\in R}$ be a C_0 group generated by A. If there exist a projection $P : X \to X$ and positive constants M, ω, μ such that:*

$$\|PT(t)\| \leq Me^{-\omega t}, \quad \forall t \geq 0, \tag{2.9}$$

$$\|(I-P)T(t)\| \leq \mu, \quad \forall t \in R, \tag{2.10}$$

then, the operator $F : X \to X$ defined by

$$Fx = \int_s^\infty (I-P)T(s-\tau)B(\tau)U(\tau,s)x d\tau, x \in X$$

is bounded and we have

$$\|F\| \leq \alpha < 1, \forall s \geq \Delta > 0.$$

Proof. Setting

$$U(t) = PT(t),$$

$$V(t) = (I-P)T(t),$$

we get

$$T(t) = U(t) + V(t).$$

By (2.1), for any $\alpha < 1$ we can find a number $\Delta > 0$ such that

$$\int_s^{+\infty} \|B(\tau)\|d\tau \leq \frac{\alpha}{\mu.K}, \forall s > \Delta > 0.$$

Lemma 2.2 implies that

$$\|F\| \leq \int_s^\infty \|V(s-\tau)\|\|B(\tau)\|\|Y(\tau,s)\|d\tau$$
$$\leq \mu.K \int_s^\infty \|B(\tau)\|d\tau \leq \alpha < 1, \forall s > \Delta > 0.$$

The Lemma 2.3 is proved. □

From Lemma 2.3 we can arrive at

Theorem 2.4. *Assume that $T(t)_{t\in R}$ is a C_0 group satisfying conditions (2.6) and (2.7) of the Lemma 2.3 . Then, $(T(t))_{t\geq 0}$ and $(Y(t,s))_{t\geq s}$ are asymptotically equivalent.*

Proof. Let $Qx = (I+F)x, x \in X$. By Lemma 2.3 we get $\|F\| < 1$. Therefore, the operator $Q: X \to X$ is invertible. Sine $U(t) = PT(t)$, $V(t) = (I-P)T(t)$ we have

$$T(t-s)V(s-\tau) = T(t-s)(I-P)T(s-\tau)$$
$$= T(t-\tau) - PT(t-\tau) = T(t-\tau) - U(t-\tau) = V(t-\tau).$$

Assume that $y(t)$ is a solution of Eq.(2.2). For each sufficiently large $s \geq 0$, $y(s) \in X$ we set $x(s) = Qy(s)$, so

$$x(s) = Qy(s) = y(s) + \int_s^\infty V(s-\tau)B(\tau)Y(\tau,s)y(s)d\tau.$$

Hence, for all $t \geq s$, we have

$$\begin{aligned} x(t) &= T(t-s)x(s) \\ &= T(t-s)y(s) + T(t-s)\int_s^\infty V(s-\tau)B(\tau)Y(\tau,s)y(s)d\tau \\ &= T(t-s)y(s) + \int_s^\infty V(t-\tau)B(\tau)Y(\tau,s)y(s)d\tau. \end{aligned}$$

Consequently

$$\|y(t) - x(t)\| = \Big\| -\int_s^\infty V(t-\tau)B(\tau)Y(\tau,s)y(s)d\tau + \int_s^t T(t-\tau)B(\tau)Y(\tau,s)y(s)d\tau \Big\|.$$

Therefore,

$$\|y(t) - x(t)\| \leq M.K\|y(s)\| \int_s^t e^{-\omega(t-\tau)}\|B(\tau)\|d\tau + \mu.K\|y(s)\| \int_t^\infty \|B(\tau)\|d\tau$$
$$\leq M_1 \int_s^t e^{-\omega(t-\tau)}\|B(\tau)\|d\tau + M_2 \int_t^\infty \|B(\tau)\|dt, \quad \forall t \geq s,$$

where $M_1 = MK\|y(s)\|$, $M_2 = \mu K\|y(s)\|$. On the other hand, for every positive number $\varepsilon > 0$, there exists a sufficiently large number t with $t > 2s$ such that the following inequalities are valid

$$\int_s^{\frac{t}{2}} e^{-\omega(t-\tau)}\|B(\tau)\|d\tau \le e^{-\frac{\omega t}{2}} \int_t^{\infty} \|B(\tau)\|d\tau < \frac{\varepsilon}{3M_1},$$

$$\int_{\frac{t}{2}}^{t} \|B(\tau)\|d\tau < \frac{\varepsilon}{3M_1},$$

$$\int_t^{\infty} \|B(\tau)\|d\tau < \frac{\varepsilon}{3M_2}.$$

Hence,

$$\|y(t) - x(t)\| \le M_1\Big(\int_s^{\frac{t}{2}} e^{-\omega(t-\tau)}\|B(\tau)\|d\tau + \int_{\frac{t}{2}}^{t} e^{-\omega(t-\tau)}\|B(\tau)\|d\tau\Big) +$$

$$+ M_2 \int_t^{\infty} \|B(\tau)\|d\tau < \frac{\varepsilon}{3} + \frac{\varepsilon}{3} + \frac{\varepsilon}{3} = \varepsilon.$$

This means that

$$\lim_{t\to\infty} \|y(t) - x(t)\| = 0.$$

By the uniqueness of solutions of Eq. (1.1) and Eq.(1.2), the map Q is a bijection between two set of solutions of Eq.(1.1) and Eq.(1.2). The theorem is proved. □

Example 2.5. In the Banach space l_∞ of bounded sequences we consider linear differential equations

$$\frac{dx(t)}{dt} = Ax(t), \tag{2.11}$$

$$\frac{dy(t)}{dt} = [A + B(t)]y(t), \tag{2.12}$$

where $t \ge 0$; $x, y \in l_\infty$, $A = diag(A_1, A_2, ..., A_n, ...)$. with

$$A_n = \begin{pmatrix} 0 & \frac{1}{n} & -1 \\ \frac{-1}{n} & 0 & \frac{1}{n} \\ 0 & 0 & -1 \end{pmatrix},$$

$B(t)$ is a linear operator on X and satisfies:

$$\int_0^{\infty} \|B(\tau)\| d\tau < \infty. \tag{2.13}$$

Assume that $(T(t))_{t\in R}$ is a C_0 group generated by A and $(Y(t,s)_{t\geq s}$ is the strongly continuous evolutionary process generated by $A + B(t)$. We can easily show that for $n \in N$,

$$T(t) = diag(T_1, T_2, ..., T_n, ...),$$

where

$$T_n = \begin{pmatrix} cos\frac{t}{n} & sin\frac{t}{n} & e^{-t} \\ -sin\frac{t}{n} & cos\frac{t}{n} & 0 \\ 0 & 0 & e^{-t} \end{pmatrix}.$$

Therefore, we can write $T(t)$ as follows:
$T(t) = U(t) + V(t) = diag(U_1(t), U_2(t), ..., U_n(t), ..) + diag(V_1(t), V_2(t), ..., V_n(t), ..)$
where

$$U_n = \begin{pmatrix} 0 & 0 & e^{-t} \\ 0 & 0 & 0 \\ 0 & 0 & e^{-t} \end{pmatrix},$$

$$V_n = \begin{pmatrix} cos\frac{t}{n} & sin\frac{t}{n} & 0 \\ -sin\frac{t}{n} & cos\frac{t}{n} & 0 \\ 0 & 0 & 0 \end{pmatrix}.$$

This shows that $\|U(t)\| \leq e^{-t}, \quad \forall t \geq 0$ and $\|V(t)\| \leq 2 < +\infty, \forall t \in \mathbb{R}$. Hence $(T(t))_{t\geq 0}$ satisfies the conditions of the Theorem 2.4. It follows that $(T(t))_{t\geq 0}$ and $(Y(t,s))_{t\geq s}$ are asymptotically equivalent.

We state below some versions of Theorem 2.4 in particular cases .

Corollary 2.6. *Let $(T(t))_{t\geq 0}$ be a immediately compact, uniformly bounded C_0 semigroup. Then, $(T(t))_{t\geq 0}$ and $(Y(t,s)_{t\geq s}$ are asymptotically equivalent.*

Proof. Since $(T(t))_{t\geq 0}$ is a immediately compact, uniformly bounded C_0 semigroup we deduce that the spectral set $\sigma(T(1))$ is countable and $\sigma(T(1)) \subset \{\lambda \in C : |\lambda| \leq 1\}$. Therefore, we have $\sigma(T(1)) = \sigma_1 \cup \sigma_2$ where $\sigma_1 \subset \{\lambda \in C : |\lambda| < 1\}, \sigma_2 \subset \{\lambda \in C : |\lambda| = 1\}$. Let P is a projection defined as follows

$$P = \frac{1}{2\pi i} \int_{\gamma} (zI - e^{A})^{-1} dz,$$

where γ is a contour enclosing σ_1 and disjoint from σ_2. It is easy to see that the conditions (2.6), (2.7) in Lemma 2.3 are satisfied. Therefore, the Corollary is proved. □

By Corollary 2.5, in particular, we obtain Levinson's theorem (see [5]):

Corollary 2.7. *If $X = R^n$ and $(T(t))_{t\geq 0}$ is a bounded C_0 semigroup on R^n, then $(T(t))_{t\geq 0}$ and $(Y(t,s)_{t\geq s}$ are asymptotically equivalent.*

3. The Asymptotic Equivalence of Linear Differential Equations in Hilbert Space

We now consider the asymptotic equivalence of solutions of differential equations in case X = H, where H is a Hilbert space. We will establish other sufficient conditions for the asymptotic equivalence of a class of triangle differential equations (see [2], p.9), for which the assumptions (2.6), (2.7) can be replaced by weaker conditions.

Let $\{e_i\}_1^\infty$ be a normalized orthogonal basis of Hilbert space H and $x = \sum\limits_{i=1}^{\infty} x_i e_i$ element of H. Then, we define projections $P_n : H \to H, n \geq 1$, as follows:

$$P_n x = \sum_{i=1}^{n} x_i e_i.$$

be a family of projections on H. We denote $H_n = \mathrm{Im} P_n$. Suppose that $J = \{n_1, n_2, \ldots, n_j, \ldots\}$ is a strictly increasing sequence of natural numbers ($n_j \to \infty$ as $j \to +\infty$).

We now consider the following linear differential equations

$$\frac{dx(t)}{dt} = Ax(t), \tag{3.14}$$

$$\frac{dy(t)}{dt} = [A + B(t)]y(t), \tag{3.15}$$

where $A \in \mathcal{L}(H)$, $B(t) \in \mathcal{L}(H), t \in [0, \infty)$ and

$$\int_0^\infty \|B(\tau)\| d\tau < \infty. \tag{3.16}$$

In the following theorem we assume that, the right hands of Eq.(3.1) and Eq.(3.2) have a triangle form; that is,

$$(A - P_m A) P_m x = 0, \tag{3.17}$$

$$(B(t) - P_m B(t)) P_m x = 0, \tag{3.18}$$

$$\forall m \in J, \ \forall x \in H.$$

Recall that $(T(t))_{t \in R}$ is a C_0 group generated by A(in Eq. (3.1)), $(Y(t,s))_{t \geq s}$ is strongly continuous evolutionary process generated by $A + B(t)$(in Eq.(3.2)), we have following

Theorem 3.1. *If $(T(t))_{t \geq 0}$ is a uniformly bounded C_0 semigroup, then $(T(t))_{t \geq 0}$ and $(Y(t,s)_{t \geq s}$ are asymptotically equivalent.*

Proof. For each $m \in J$ we denote by $(X_m(t))_{t \geq 0}$ a C_0 semigroup generated by AP_m , and by $(Y_m(t,s))_{t \geq s}$ a strongly continuous evolutionary process generated by $(A + B(t))P_m$. By the assumption, $(T(t))_{t \geq 0}$ is uniformly bounded, we see that $(X_m(t))_{t \geq 0}$ is a immediately compact,bounded C_0 semigroup in H_m. By the Corollary 2.6, we show that, there exist a projection $\Pi_m : H_m \to H_m$ and positive constants a_m, b_m, c_m such that

$$\|\Pi_m X_m(t)\| \leq a_m e^{-b_m t}, \forall t \geq 0, \tag{3.19}$$

$$\|(I_m - \Pi_m)X_m(t)\| \le c_m, \quad \forall t \in R, \tag{3.20}$$

for all $m \in J$.
Let

$$F_m = \int_s^\infty (I_m - \Pi_m)X_m(t)B(\tau)Y_m(t,\tau)d\tau. \tag{3.21}$$

Using Lemma 2.3 we show that there exists a positive number $\Delta = \Delta(\alpha)$ (independent of m) such that

$$\|F_m\| \le \alpha < 1, \quad \forall s \ge \Delta, \ \forall m \in J.$$

We now prove that $\{F_m\}$ is a convergent sequence of operators as $m \to \infty$. Indeed, by (3.4) and (3.5) we obtain that for all $m, p \in J$:

$$X_{m+p}(t-s)P_m\xi = X_m(t-s)P_m\xi, \forall \xi \in H,$$

$$Y_{m+p}(t,s)P_m\xi = Y_m(t,s)P_m\xi, \forall \xi \in H.$$

Hence,

$$F_{m+p}P_m\xi = F_mP_m\xi, \forall \xi \in H.$$

The convergence of $\{F_m\}$ follows from the estimate

$$\begin{aligned}\|F_{m+p} - F_m\| &= \|F_{m+p}P_{m+p} - F_mP_m\| \\ &= \|F_{m+p}(P_{m+p} - P_m) + (F_{m+p} - F_m)P_m\| \\ &= \|F_{m+p}(P_{m+p} - P_m)\| \\ &\le \|F_{m+p}\|\|P_{m+p} - P_m\|.\end{aligned}$$

for all $m, p \in J$. Since the sequence $\{P_m\}$ is convergent and the sequence $\{F_m\}$ is uniformly bounded, we have that $\{F_m\}$ is a convergent sequence. Let $Q_m x = (I + F_m)x, x \in H_m$. We now put:

$$F = \lim_{m\to\infty} F_m \quad \text{and} \quad Q = \lim_{m\to\infty} Q_m.$$

Then, $Q = I + F$. Since $\|F_m\| \le \alpha < 1, \forall m \in J, \forall s \ge \Delta$, we have

$$\|F\| \le \alpha < 1, \quad \forall s \ge \Delta.$$

By the invertibility of $Q : H \to H$ and uniqueness of solutions of Eq.(3.1) and Eq.(3.2) we deduce that the map Q is also a bijection between two sets of solutions $\{x(t)\}$ of Eq.(3.1) and $\{y(t)\}$ of Eq.(3.2). Put $y_0 = Q^{-1}x_0$ and consider $x(t) = T(t-s)x_0$ and $y(t) = Y(t,s)y_0$. Then we have

$$\lim_{m\to\infty} P_m y_0 = y_0, \quad \lim_{m\to\infty} Q_m y_0 = Qy_0 = x_0.$$

Denote $x(t; s, x_0) = T(t-s)x_0$, $y(t; s, y_0) = Y(t, s)y_0$. We can deduce that for any given $\varepsilon > 0$, there exists $m_1 \in J$ sufficiently large such that for all $m \geq m_1$ we have

$$\|y(t; s, y_0) - y(t; s, P_m y_0)\| < \frac{\varepsilon}{3},$$
$$\|x(t; s, y_0) - x(t; s, Q_m y_0)\| < \frac{\varepsilon}{3}, \text{ for all} t \geq s.$$

On the other hand, Theorem 2.4 implies $(X_m(t))_{t\geq 0}$ and $(Y_m(t, s))_{t\geq s}$ are asymptotically equivalent. Therefore, there exists $\tau_0 \in (s, \infty)$ such that for all $t \geq \tau_0$ we have

$$\|x(t; s, Q_{m_1} y_0) - y(t; s, P_{m_1} y_0)\| < \frac{\varepsilon}{3},$$

thus,

$$\begin{aligned}\|y(t; s, y_0) - x(t; s, x_0)\| &\leq \|y(t; s, y_0) - y(t; s, P_{m_1} y_0)\| + \\ &+ \|y(t, s, P_{m_1} y_0) - x(t, s, Q_{m_1} y_0)\| \\ &+ \|x(t; s, Q_{m_1} y_0) - x(t; s, x_0)\| \\ &\leq \frac{\varepsilon}{3} + \frac{\varepsilon}{3} + \frac{\varepsilon}{3} = \varepsilon, \quad \forall t \geq \tau_0 \geq s,\end{aligned}$$

for each sufficiently large $s \geq 0$. This implies that

$$\lim_{t\to\infty} \|x(t; s, x_0) - y(t; s, y_0)\| = 0.$$

It means $(T(t))_{t\geq 0}$ and $Y(t, s)_{t\geq s}$ are asymptotically equivalent, so the theorem is proved. □

By the same methods as used in proof of Theorem 3.1, we obtain the following corollary.

Corollary 3.2. *Assume that $A \in L(H)$ is a compact self-adjoint operator and $(T(t))_{t\geq 0}$ is a bounded uniformly C_0 Semigroup then $(T(t))_{t\geq 0}$ and $Y(t, s)_{t\geq s}$ are asymptotically equivalent.*

References

[1] R. Nagel (Ed.), *One-Parameter Semigroups of Positive Operators*, Lecture Notes in Math. vol. **1184**, Springer-Verlag, Berlin, 1986.

[2] Dang Dinh Chau, On the asymptotic equivalence of linear differential equations in Hilbert spaces, *Vietnam National University-Hanoi Journal of Science, Math.-Phys. Series* (2002), No. 2, 8-17.

[3] J. Kato, The asymptotic equivalence of functional of differential equations, *J. Diff. Eq.* **1** (1965), 306–332.

[4] X.G. Krein, *Linear equation differential in Banach space*, "Nayka" Moscow, 1967.

[5] N.Levinson, The asymptotic behavior of systems of linear differental equations *Amer. J. Math*, **63** (1946),1-6.

[6] J. Miklo, Asymptotic relationship between solution of two linear diffirential systems, *Bratislava Mathematica Bohemica* **123** (1998), 163-175.

[7] T. Naito, Nguyen Van Minh and J. S. Shin, New Sectral Criteria for Almost Periodic Soutions of Evolution Equations. *Studia Mathematica* **145** (2001),97-111.

[8] Nguyen The Hoan, Some asymptotic behaviors of solutions to non-linear system of differential equations, *Diff. Eq.* **12** (1981), 385-360 (Russian).

[9] A. Pazy, *Semigroups of Linear Operators and Applications to Partial Diffirential Equations*, Springer-Verlag, Berlin- New York 1983.

[10] E.V. Voskoresenski, *Asymptotic equivalence of systems of differential equations,* Progress of Mathematic Sciences, **5**(1985), C.40,N_0 2 (1985), (*Ukrainian J. Math.*, **40** (1985), 249-250).(Russian)

In: Focus on Evolution Equations
Editor: Gaston M. N'Guerekata

ISBN: 978-1-60021-342-7

Chapter 10

ON SUPERPOSITION OF ALMOST PERIODIC FUNCTIONS

Stanisław Stoiński[*]
Faculty of Mathematics and Computer Science,
Adam Mickiewicz University,
Umultowska 87, 61-614 Poznań

Abstract

In this note we present some theorem on the superposition of an H-almost periodic (a.p.) function and the reciprocal of trigonometric polynomial of constant sign. Moreover, we prove theorems on the superposition of a continuous function and an H-a.p. function, as well an N-a.p. function. Finally, we prove a theorem on the superposition of a differentiable function and a Bohr's a.p. function (B-a.p.).

1991 Mathematics Subject Classification: 42A75

Keywords and phrases: Almost periodic function, superposition, Hausdorff metric, variation

1. Introduction

The continuous functions which are almost periodic (a.p. for short) with respect to the uniform metric was defined and studied by H. Bohr (see [2]) in 1925 and later. V.V. Stepanov, H. Weyl and A.S. Besicovitch had studied the properties of functions a.p. with respect to integral metrics (see [1], [3], [6]). B.M. Levitan had considered some kind of generalization of Bohr's a.p. function, namely N-a.p. functions (see [6]). However functions a.p. with respect to the Hausdorff metric (see [5]) (which was studied by B. Sendov and others , see [7]) , so-called H-a.p. functions was defined and investigated by S. Stoiński (see [8]-[11]) and a little later by A.S. Djafarov and G.M. Gasanov (see [4]).

It is well known (see [6], p.21) that the superposition of uniformly continuous function and Bohr's a.p. function is Bohr's a.p. function. The superpositions of different a.p. functions and continuous or uniformly continuous functions, was considered by S. Stoiński in

[*]E-mail address: stoi@amu.edu.pl

[9] and [13]. This paper is concerned with the superposition of such functions as: Bohr's a.p. or N-a.p. and H-a.p. functions and continuous or differentiable functions. The results of this type have application in the field of differential equations.

2. Preliminaries

Let F_Δ denote the class of closed and bounded with respect to the y axis point sets on the Oxy plane, such that their respective projections on the x axis are identical with the interval Δ (bounded or unbounded), and such that the intersection of every straight line $x = x_0$, $x_0 \in \Delta$, and $F \in F_\Delta$ is a closed interval. If $A, B \in F_\Delta$, then the Hausdorff distance between A and B we call the following number

$$r_\Delta(A, B) = \max(\sup_{X \in A} \inf_{Y \in B} \|X - Y\|_0, \sup_{X \in B} \inf_{Y \in A} \|X - Y\|_0),$$

where

$$\|X - Y\|_0 = \|X(x_1, y_1) - Y(x_2, y_2)\|_0 = \max(|x_1 - x_2|, |y_1 - y_2|)).$$

The functional $r_\Delta : F_\Delta \times F_\Delta \to \mathbb{R}_+$ is a metric in F_Δ. The following lemma is true (see [7]):

Lemma 2.1. *Suppose that $A, B \in F_\Delta$. If:*
(a) for an arbitrary $X \in A$ there exists $Y \in B$ such that $\|X - Y\|_0 \leq \delta$
and
(b) for an arbitrary $X \in B$ there exists $Y \in A$ such that $\|X - Y\|_0 \leq \delta$,
then $r_\Delta(A, B) \leq \delta$.

By the complete graph $\bar{f}$ of the function f, defined and bounded on the interval Δ and taking real values, we call a set of points (x, y), for which $x \in \Delta$, $I_f(x) \leq y \leq S_f(x)$, where I_f, S_f is the lower and upper Baire's function respectively, both with respect to the function f:

$$I_f(x) = \lim_{\delta \to 0} \inf_{|x' - x| \leq \delta} f(x') \qquad S_f(x) = \lim_{\delta \to 0} \sup_{|x' - x| \leq \delta} f(x').$$

The Hausdorff distance between two functions f and g, both defined and bounded on Δ, is defined in the following way: it is the Hausdorff distance between their respective complete graphs, i.e. $r_\Delta(f, g) = r_\Delta(\bar{f}, \bar{g})$. Let $r_\mathbb{R} = r$.

A set $E \subset \mathbb{R}$ is called relatively dense iff there exists a positive number l such that in every open interval $(\omega, \omega + l)$, $\omega \in \mathbb{R}$, there is at least one element of the set E.

Let $f \colon \mathbb{R} \to \mathbb{R}$ be a bounded function. If for $\varepsilon > 0$ there is $r(f_\tau, f) \leq \varepsilon$, where $f_\tau(x) \equiv f(x + \tau)$,then the number $\tau \in \mathbb{R}$ is called an (H, ε)- almost period ((H, ε) -a.p.) of the function f. Let us denote the set of (H, ε)-a.p. of f by $E_H\{\varepsilon; f\}$. A bounded function $f \colon \mathbb{R} \to \mathbb{R}$ is called H-almost periodic (H-a.p.) iff for each $\varepsilon > 0$ the set $E_H\{\varepsilon; f\}$ is relatively dense. Let $\widetilde{H}$ be the space of H-a.p. functions. Examples, elementary properties of H-a.p. functions and the connection between H-a.p. functions and almost periodic functions of other types can be found in [4] and [8]-[11].

It is known (see [11]), that if F is a Bohr's a.p. function, i.e. a uniformly a.p. function (see [6]), f is a trigonometric polynomial of constant sign, then the superposition $F \circ f^{-1}$ $((F \circ f^{-1})(x) \equiv F(1/f(x)))$ is H-a.p.

If f is a uniformly continuous function on the set Y_g of values of the function g, where g is a continuous H-a.p. function, then $f \circ g$ is H-a.p. This was proved by me in [9].

Let $V(f;x)$ denote the Jordan variation of a function $f\colon \mathbb{R} \to \mathbb{R}$ on the interval $\langle x - 1, x+1\rangle$ for an arbitrary $x \in \mathbb{R}$ and let

$$V(f) = \sup\{|f(x)| + V(f;x) : \ x \in \mathbb{R}\}.$$

Let us put

$$BV_{loc} = \{f\colon \mathbb{R} \to \mathbb{R} : \ V(f;x) < \infty \text{ for every } x \in \mathbb{R}\}.$$

A continuous function $f \in BV_{loc}$ is called almost periodic in variation (V-a.p.) iff for each $\varepsilon > 0$ the set $\{\tau \in \mathbb{R} : \ V(f - f_\tau) \leq \varepsilon\}$ is relatively dense. Elementary properties of functions almost periodic in variation can be found in [12].

Let $f\colon \mathbb{R} \to \mathbb{R}$ be a continuous function. If for $N > 0, \varepsilon > 0$ there is

$$D_N(f - f_{\pm\tau}) = \max\{|f(x) - f_{\pm\tau}(x)| : -N \leq x \leq N\} \leq \varepsilon,$$

then the number $\tau \in \mathbb{R}$ is called an (N, ε)-almost period ((N,ε)-a.p.) of the function f. Let $NE\{\varepsilon; f\}$ denote the set of (N,ε)-almost periods of f. A continuous function $f\colon \mathbb{R} \to \mathbb{R}$ is called N-almost periodic (N-a.p.) iff there exists a uniformly a.p. function φ, a dominant function of f, such that for any two positive numbers N, ε there exists a $\delta > 0$ such that $E\{\delta; \varphi\} \subset NE\{\varepsilon; f\}$, where $E\{\delta;\varphi\} = \{\tau \in \mathbb{R} : \sup\{|\varphi(x) - \varphi_\tau(x)| : x \in \mathbb{R}\} \leq \varepsilon\}$. By $\tilde{N}$ we denote the space of N-a.p. functions. The above definitions were introduced by B. M. Levitan (see [6], pp.141-145).

It is known (see [13]) that if g is a Bohr's a.p. function and a function $f\colon Y_g \to \mathbb{R}$ is continuous, then the superposition $f \circ g$ is N-a.p.

3. Main Results

Now, we present the theorem on the superposition of an H-a.p. function and the reciprocal of trigonometric polynomial of constant sign. In addition, we prove theorems on the superposition of a continuous function and an H-a.p. function, as well an N-a.p. function. Moreover, we prove the theorem on the superposition of a differentiable function and a B-a.p. function.

Theorem 3.1. *If F is a continuous H-a.p. function, f is a trigonometric polynomial of constant sign, then the superposition $F \circ f^{-1}$ is H-a.p.*

Proof. Let us assume that $f(x) > 0$ for every $x \in \mathbb{R}$ and $\inf\{f(x) : x \in \mathbb{R}\} = 0$. We proceed in the same way if $f(x) < 0$ for every $x \in \mathbb{R}$.

A Bohr's a.p. function f is bounded ([6], p.22), and so there exists a positive constant M such that $0 < f(x) \leq M$ for every $x \in \mathbb{R}$. Let us denote $E_\alpha = \{x \in \mathbb{R} : f(x) > \alpha\}$ for $0 < \alpha < M$. It is known (see [6], pp.210-211) that for an arbitrary $0 < \varepsilon < 1/2$ there exists $\alpha_0 > 0$ such that for every $x \in \mathbb{R}$ we have $\mu(CE_\alpha(x, x+1)) < \varepsilon/2$ for $0 < \alpha \leq \alpha_0$, where $CE_\alpha(x, x+1)) = \{t \in \langle x, x+1\rangle : f(t) \leq \alpha\}$ for $x \in \mathbb{R}$ and μ is the Lebesgue measure.

Because F is an H-a.p. function, the following equality

$$F(u) = F_{\bar{\tau}}(u + \vartheta_1 \varepsilon) + \vartheta_2 \varepsilon, \quad u \in \mathbb{R}, \tag{3.1}$$

for $\bar{\tau} \in E_H\{\varepsilon; F\}$, where $|\vartheta_i| = |\vartheta_i(u, \bar{\tau})| \leq 1$, $i = 1, 2$, holds. Let $\bar{l} = \bar{l}(\varepsilon) > 0$ be a number which characterizes the relative density of the set $E_H\{\varepsilon; F\}$.

In the following we assume that for an arbitrary $0 < \varepsilon < 1/2$ and $\bar{l} > 1$

$$n_0 > \max\left(\frac{1}{\alpha_0 M}, \frac{1 + 12\bar{l}\alpha_0}{\alpha_0^2}\right)$$

$$\frac{1}{\alpha_0 n_0 - 12\bar{l}} < \alpha \leq \alpha_0, \qquad \frac{1}{\alpha_0 n_0} < \alpha' \leq \frac{\alpha}{1 + 12\alpha\bar{l}}$$

and

$$0 < \delta < \min\left(\alpha' - \frac{1}{n_0\alpha_0}, \frac{\alpha'^2\Delta}{1 + \alpha'\Delta}, \frac{4\alpha'^2\bar{l}}{1 - 4\alpha'\bar{l}}, \frac{4\alpha^2\bar{l}}{1 + 4\alpha\bar{l}}\right),$$

where $\Delta = \Delta(\varepsilon) > 0$ is a number which characterizes the uniform continuity of a function F on the closed interval $\langle 1/M, \alpha_0 n_0\rangle$.

For $x \in E_{\alpha'}$ and $\tau \in E\{\delta; f\}$ the following inequality

$$\left|\frac{1}{f(x+\tau)} - \frac{1}{f(x)}\right| < \frac{\delta}{(\alpha' - \delta)\alpha'}$$

holds, and so

$$\left|F\left(\frac{1}{f(x+\tau)}\right) - F\left(\frac{1}{f(x)}\right)\right| < \varepsilon. \tag{3.2}$$

Writing $CE_\alpha = \{x \in \mathbb{R} : f(x) \leq \alpha\}$ we obtain

$$CE_\alpha = \bigcup_{k=-\infty}^{\infty} CE_\alpha(k, k+1) = \bigcup_{k=-\infty}^{\infty} \omega_k.$$

It is easily seen that for $x \in CE_{\alpha'} = \bigcup_{k=-\infty}^{\infty} \omega'_k$, i.e. for $x \in \omega'_k \neq \emptyset$, $k = 0, \pm 1, \pm 2, \ldots$, and $\tau \in E\{\delta; f\}$ the following estimations

$$\frac{1}{f(x)} \geq \frac{1}{\alpha} + 12\bar{l}, \tag{3.3}$$

$$\frac{1}{f_\tau(x)} \geq \frac{1}{\alpha} + 8\bar{l}, \tag{3.4}$$

$$\frac{1}{f_\tau(x)} > \frac{1}{\alpha - \delta} + 4\bar{l} \tag{3.5}$$

hold.

Now, we present the estimate of the Hausdorff distance

$$r\big((F\circ f^{-1})_\tau,(F\circ f^{-1})\big)$$

where $\tau \in E\{\delta; f\}$.

We prove that for every $X = X\big(x, F(1/f_\tau(x))\big)$, where $x \in CE_{\alpha'}$, there exists $Y = Y\big(x+\vartheta\varepsilon, F(1/f(x+\vartheta\varepsilon))\big)$ such that $\|X-Y\|_0 \le \varepsilon$, where $|\vartheta| < 1$. Indeed, if $1/\alpha+8\bar{l} \le 1/f_\tau(x) \le 1/\alpha+12\bar{l}$, then in the interval $(x-\varepsilon, x+\varepsilon)$ there exists $x' \notin CE_\alpha$, i.e. $1/f(x') < 1/\alpha$. A function f is continuous on the closed interval $\langle x, x'\rangle$. Hence, using (3.3) and (3.4), there can be found a ϑ , $|\vartheta| < 1$, such that $1/f(x+\vartheta\varepsilon) = 1/f_\tau(x)$.

If $1/f_\tau(x) > 1/\alpha+12\bar{l}$, then we can find an integer k such that $1/f_\tau(x)+k\bar{l} > 1/\alpha+12\bar{l}$ and $1/f_\tau(x)+(k-1)\bar{l} \le 1/\alpha+12\bar{l}$. For $\bar{\tau} \in ((k-3)\bar{l},(k-2)\bar{l})$ the following inequality $1/f_\tau(x)+\bar{\tau}+\vartheta_1(x,\bar{\tau})\varepsilon < 1/\alpha+12\bar{l}$ holds. Because $\mu(\omega_k) < \varepsilon/2$, so in the interval $(x-\varepsilon, x+\varepsilon)$ there exists $x'' \neq CE_\alpha$, i.e. $1/f(x'') < 1/\alpha$. A function f is continuous on the closed interval $\langle x, x''\rangle$. From (3.3) it follows that we can find a ϑ, $|\vartheta| < 1$, such that $1/f(x+\vartheta\varepsilon) = 1/f_\tau(x)+\bar{\tau}+\vartheta_1\varepsilon$. Substituting $u = 1/f_\tau(x)$ into (3.1), we obtain $|F(1/f_\tau(x)) - F(1/f(x+\vartheta\varepsilon))| \le \varepsilon$.

In the following we prove that for every $X = X(x, F(1/f(x)))$, where $x \in CE_{\alpha'}$, there exists $Y = Y\big(x+\vartheta\varepsilon, F(1/f_\tau(x+\vartheta\varepsilon))\big)$ such that $\|X-Y\|_0 \le \varepsilon$, where $|\vartheta| < 1$.

If $1/f(x) \ge 1/\alpha+12\bar{l}$, and so $1/f(x) \ge 1/(\alpha-\delta)+4\bar{l}$, then we can find an integer k such that $1/f(x)+k\bar{l} \ge 1/(\alpha-\delta)+4\bar{l}$ and $1/f(x)+(k-1)\bar{l} < 1/(\alpha-\delta)+4\bar{l}$. For $\bar{\tau} \in ((k-3)\bar{l},(k-2)\bar{l})$ the following inequality $1/f(x)+\bar{\tau} < 1/(\alpha-\delta)+4\bar{l}$ holds. Because $\mu(\omega_k) < \varepsilon/2$, so in the interval $(x-\varepsilon, x+\varepsilon)$ there exists $x''' \notin CE_\alpha$, i.e. $f(x''') > \alpha$, and $1/f_\tau(x''') < 1/(\alpha-\delta)$. A function f_τ is continuous on the closed interval $\langle x, x'''\rangle$. From (3.5) it follows that we can find a ϑ , $|\vartheta| < 1$, such that $1/f_\tau(x+\vartheta\varepsilon) = 1/f(x)+\bar{\tau}+\vartheta_1\varepsilon$. Substituting $u = 1/f(x)$ into (3.1) , we obtain $|F(1/f(x)) - F(1/f_\tau(x+\vartheta\varepsilon))| \le \varepsilon$.

Thanks to the arbitrarily chosen ω'_k, $k = 0, \pm1, \pm2, \ldots$, and from (3.2) we get by Lemma 2.1 $r\big((F\circ f^{-1})_\tau, F\circ f^{-1}\big) < \varepsilon$ for $\tau \in E\{\delta, f\}$. Therefore, the superposition $F\circ f^{-1}$ is an H-a.p. function.

Now, let $\inf\{f(x) : x \in \mathbb{R}\} = m > 0$. Since F is uniformly continuous on the closed interval $\langle 1/M; 1/m\rangle$, so the superposition $F\circ f^{-1}$ is a Bohr's a.p. function. The proof is complete. □

The number $\tau \in \mathbb{R}$ is called an (NH,ε)-almost period ((NH,ε)-a.p.) of a bounded on every closed interval function $f\colon \mathbb{R} \to \mathbb{R}$ iff for two positive numbers N, ε we have $r_{\langle -N,N\rangle}(f, f_\tau) \le \varepsilon$. Let us denote by $NE_H\{\varepsilon; f\}$ the set of (NH,ε)-almost periods of f.

A bounded on every closed interval function $f\colon \mathbb{R} \to \mathbb{R}$ is called (NH)-almost periodic ((NH)-a.p.) iff there exists an H-a.p. function h, a dominant function of f, such that for any two positive numbers N, ε there exists a $\delta = \delta(N,\varepsilon) > 0$ such that $E_H\{\delta, h\} \subset NE_H\{\varepsilon; f\}$.

Theorem 3.2. *If f is a continuous function on the set Y_g, where g is a continuous H-a.p. function, then the superposition $f\circ g$ is (NH)-a.p.*

Proof. It is known (see [9]) that g is bounded on $\mathbb{R}$. Let us denote $m = \inf\{g(x) : x \in \mathbb{R}\}$, $M = \sup\{g(x) : x \in \mathbb{R}\}$. Fix arbitrarily $N > 0$, $\varepsilon > 0$. Let us write $m_N = \min\{g(x) :$

$-N \le x \le N\}$, $M_N = \max\{g(x) : -N \le x \le N\}$ and

$$I_N = \begin{cases} \langle m_N - a_N, M_N + a_N \rangle, & \text{where } 0 < a_N = \frac{1}{2}\min\{m_N - m, M - M_N\}, \\ & \text{if } m_N > m,\ M_N < M, \\ \langle m_N, M_N + a_N \rangle, & \text{where } 0 < a_N = \frac{1}{2}(M - M_N), \\ & \text{if } m_N = m,\ M_N < M, \\ \langle m_N - a_N, M_N \rangle, & \text{where } 0 < a_N = \frac{1}{2}(m_N - m), \\ & \text{if } m_N > m,\ M_N = M. \end{cases}$$

If $m = m_N$, $M = M_N$, then f is a uniformly continuous function on the set Y_g, and so the superposition $f \circ g$ is H-a.p. (see [9]).

Let $0 < \delta = \delta(N, \varepsilon/2) \le a_N$ be a number which characterizes the uniform continuity of the function f on the closed interval I_N and $0 < \delta' < \min(\varepsilon/2, \delta, N)$.

Because g is a continuous H-a.p. function , the following equality $g_\tau(x) = g(x + \vartheta_1(x,\tau)\delta') + \vartheta_2(x,\tau)\delta'$ for $\tau \in E_H\{\delta'; g\}$, $x \in \mathbb{R}$, where $|\vartheta_i(x,\tau)| \le 1$, $i = 1, 2$, holds. Hence, for every $X \in A' = \{(x,y) : x \in \langle -N + \delta', N - \delta' \rangle,\ y = f(g_\tau(x))\}$ there exists $Y = Y(x + \vartheta_1(x,\tau)\delta', f(g(x + \vartheta_1(x,\tau)\delta'))) \in \overline{f \circ g}$ such that $\|X - Y\|_0 < \varepsilon/2$. By the equality $g(x) = g_\tau(x + \vartheta_1(x,\tau)\delta') + \vartheta_2(x,\tau)\delta'$ for $\tau \in E_H\{\delta'; g\}$, $x \in \mathbb{R}$, where $|\vartheta_i(x,\tau)| \le 1$, $i = 1, 2$, it follows that for every $X \in B' = \{(x,y) : x \in \langle -N + \delta', N - \delta' \rangle,\ y = f(g(x))\}$ there exists $Y = Y(x + \vartheta_1(x,\tau)\delta', f(g_\tau(x + \vartheta_1(x,\tau)\delta'))) \in \overline{(f \circ g)_\tau}$ such that $\|X - Y\|_0 < \varepsilon/2$. By the uniform continuity of a function f on the interval I_N it follows that for every $X \in A = \{(x,y) :\ x \in \langle -N, N \rangle,\ y = f(g_\tau(x))\}$ there exists $Y_0 \in A'$ such that $\|X - Y_0\|_0 \le \varepsilon/2$, and so for every $X \in A$ there exists $Y \in B = \{(x,y) :\ x \in \langle -N, N \rangle,\ y = f(g(x))\}$ such that $\|X - Y\|_0 < \varepsilon$. Analogously, we obtain that for every $X \in B$ there exists $Y \in A$ such that $\|X - Y\|_0 < \varepsilon$. Therefore, from the above estimates we get by Lemma 2.1

$$r_{\langle -N,N \rangle}((f \circ g)_\tau, f \circ g) < \varepsilon,$$

and so the superposition $f \circ g$ is an (NH)-a.p. function, because every (H, δ')-a.p. of g is an (NH, ε)-a.p. of $f \circ g$, i.e. g is a dominant function of $f \circ g$. □

Example 3.3. Let f be a B-a.p. function such that $f(x) > 0$ for every $x \in \mathbb{R}$ and $\inf\{f(x) : x \in \mathbb{R}\} = 0$, (for example : $f(x) = 2 + \cos x + \cos(\sqrt{2}x)$ for $x \in \mathbb{R}$ (see [6], p.144)). Moreover, let g be a trigonometric polynomial of constant sign. Then

$$h(x) = \ln\Big(f\big(\frac{1}{g(x)}\big)\Big)$$

for $x \in \mathbb{R}$ is the unbounded (NH)-a.p. function, if only we choose adequately the polynomial g. It suffice to take $g(x) = f(x) = 2 + \cos x + \cos(\sqrt{2}x)$ for $x \in \mathbb{R}$.

Now, we will prove the following theorem:

Theorem 3.4. *If f is a continuous function on the set Y_g, where g is an N-a.p. function, then the superposition $f \circ g$ is N-a.p.*

Proof. Fix arbitrarily $N > 0$, $\varepsilon > 0$. Let us write $m_N = \min\{g(x) : -N \leq x \leq N\}$, $M_N = \max\{g(x) : -N \leq x \leq N\}$. By the Darboux property it follows that $\{g(x) : -N \leq x \leq N\} = \langle m_N, M_N \rangle$.

Because g is N-a.p. , there exists a B-a.p. function φ , a dominant function of g, such that for any two positive numbers N, ε' there exists a $\delta' = \delta'(N, \varepsilon') > 0$ such that $E\{\delta'; \varphi\} \subset NE\{\varepsilon'; g\}$. Hence, for every $x \in \langle -N, N \rangle$ and $\tau \in E\{\delta'; \varphi\}$ we have

$$|g(x+\tau) - g(x)| \leq \varepsilon'. \tag{3.6}$$

If $Y_g = \langle m, M \rangle$, where $m = \inf\{g(x) : x \in \mathbb{R}\}$, $M = \sup\{g(x) : x \in \mathbb{R}\}$, then f is uniformly continuous on the closed interval Y_g. Substituting $\varepsilon' = \delta = \delta(\varepsilon) < \Delta = \Delta(\varepsilon)$ into (3.6) , where Δ is a number which characterizes the uniform continuity of the function f on Y_g, we obtain that for every $x \in \langle -N, N \rangle$ and $\tau \in E\{\delta'; \varphi\}$ the following estimates

$$|g(x+\tau) - g(x)| \leq \delta, \qquad |f(g(x+\tau)) - f(g(x))| < \varepsilon$$

hold, i.e. $\tau \in NE\{\varepsilon; f \circ g\}$, and so φ is a dominant function of $f \circ g$.

In the following we assume that $\langle m_N, M_N \rangle \subsetneq Y_g$. If Y_g is a bounded interval, then we construct the closed interval

$$I_N = \begin{cases} \langle m_N - a_N, M_N + a_N \rangle, & \text{where} \quad 0 < a_N = \frac{1}{2}\min(m_N - m, M - M_N), \\ & \text{if} \quad m_N > m, \ M_N < M, \\ \langle m_N, M_N + a_N \rangle, & \text{where} \quad 0 < a_N = \frac{1}{2}(M - M_N), \\ & \text{if} \quad m_N = m, \ M_N < M, \\ \langle m_N - a_N, M_N \rangle, & \text{where} \quad 0 < a_N = \frac{1}{2}(m_N - m), \\ & \text{if} \quad m_N > m, \ M_N = M. \end{cases}$$

In case when Y_g is an unbounded interval, then we construct the closed interval

$$I'_N = \begin{cases} \langle m_N - 1, M_N + 1 \rangle, & \text{if} \quad Y_g = \mathbb{R}, \\ \langle m_N - 1, M_N + a_N \rangle, & \text{where} \quad 0 < a_N = \frac{1}{2}(M - M_N), \\ & \text{if} \quad Y_g = (-\infty, M), \\ \langle m_N - 1, M_N \rangle, & \text{if} \quad M_N < M, \\ \langle m_N - a_N, M_N + 1 \rangle, & \text{where} \quad 0 < a_N = \frac{1}{2}(m_N - m), \\ & \text{if} \quad m_N > m, \\ \langle m_N, M_N + 1 \rangle, & \text{if} \quad m_N = m. \end{cases}$$

Let $0 < \Delta' = \Delta'(N, \varepsilon)$ be a number which characterizes the uniform continuity of a function f on the interval I_N or I'_N. In the case of the interval I_N or when the interval I'_N is defined by a_N we assume that $\Delta' \leq \min(1, a_N)$. In the other cases we assume that $\Delta' \leq 1$. Substituting $\varepsilon' = \delta = \delta(N, \varepsilon) < \Delta'$ into (3.6), we have that for every $x \in \langle -N, N \rangle$ and $\tau \in E\{\delta', \varphi\}$

$$|g(x+\tau) - g(x)| \leq \delta \qquad \text{and} \qquad |f(g(x+\tau)) - f(g(x))| < \varepsilon,$$

and so $\tau \in NE\{\varepsilon, f \circ g\}$, i.e. φ is dominant function of $f \circ g$.

Since $E\{\delta'; \varphi\}$ is a symmetric set, the inequality $|f(g(x-\tau)) - f(g(x))| < \varepsilon$ for every $x \in \langle -N, N \rangle$, holds too.

Hence, the superposition $f \circ g$ is an N-a.p. function. □

The following theorem, which concerns V-a.p. functions (see [12]), is true:

Theorem 3.5. *If the derivative f' of a function $f\colon Y_g \to \mathbb{R}$ is uniformly continuous and bounded on Y_g, where g is a B-a.p. function such that the derivative g' is uniformly continuous, then the superposition $f \circ g$ is V-a.p.*

Proof. Let us write
$$M_{f'} = \sup\{|f'(u)| : u \in Y_g\}.$$

For an arbitrary $\varepsilon > 0$ and for every $u \in Y_g$ we have
$$|f(u+h) - f(u)| = |f'(u+\vartheta h)||h| \le \big(|f'(u+\vartheta h) - f'(u)| + |f'(u)|\big)|h| < \varepsilon$$
for $|h| < \delta = \min(\delta_{f'}, \varepsilon/(\varepsilon + M_{f'}))$, where $0 < \vartheta < 1$, $\delta_{f'} = \delta_{f'}(\varepsilon) > 0$ is a number which characterizes the uniform continuity of a function f' on Y_g, and so f is uniformly continuous on Y_g. Applying the theorem on the superposition of a uniformly continuous function and a B-a.p. function (see [6], p.21), we conclude that $f \circ g$ is B-a.p.

Since f' is bounded on Y_g and g' is continuous, so the variation of $f \circ g$ on every closed interval is finite.

The derivative $(f \circ g)'$ is uniformly continuous on $\mathbb{R}$, when for an arbitrary $\varepsilon > 0$ and for every $x \in \mathbb{R}$ we obtain
$$|(f \circ g)'(x+h) - (f \circ g)'(x)|$$
$$\le |f'(g(x+h)) - f'(g(x))||g'(x+h)| + |f'(g(x))||g'(x+h) - g'(x)| < \varepsilon$$
for $|h| < \delta = \min(\delta_g, \delta_{g'})$, where δ_g, $\delta_{g'}$ are the numbers which characterizes the uniform continuity of g and g' on $\mathbb{R}$.

It is known (see [12]) that if the derivative of a B-a.p. function is uniformly continuous, then this function is V-a.p. Hence, the superposition $f \circ g$ is a.p. in variation. This finishes the proof. □

Acknowledgment

I would like to express my gratitude to the Reviewer for his invaluable critical remarks and suggestions.

References

[1] A. S. Besicovitch, *Almost periodic functions*, Cambridge 1932.

[2] H. Bohr, Zur Theorie der fastperiodischen Funktionen, I Teil, *Acta Math.* **45** (1925), 29–127; II Teil, *Acta Math.* **46** (1925), 101–214.

[3] C. Corduneanu, *Almost Periodic Functions*, Chelsea Publishing Co., New York, 1989.

[4] A. S. Djafarov and G. M. Gasanov, Almost periodic functions of the relatively Hausdorff metric and some of their properties, *Izv. Acad. Azerb. Sci. Ser. Sci. Phys. Tech. Math.* **1** (1977), 57–62 (in Russian).

[5] F. Hausdorff, *Grundzüge der Mengenlehre*, Leipzig 1914.

[6] B. M. Levitan, *Almost periodic functions*, Moscow 1953 (in Russian).

[7] B. Sendov and B. Penkov, ε-entropy and ε-capacity of the space of continuous functions, *Izv. Math. Inst. Acad. Bulg. Sci.* **6** (1962), 27–50 (in Bulgarian).

[8] S. Stoiński, H-almost periodic functions, *Funct. Approximatio Comment. Math.* **1** (1974), 113–122.

[9] S. Stoiński, A connection between H-almost periodic functions and almost periodic functions of other types, *Funct. Approximatio Comment. Math.* **3** (1976), 205–223.

[10] S. Stoiński, On modulus of non-monotonicity of an H-almost periodic function, *Funct. Approximation Comment. Math.* **4** (1976), 85–90.

[11] S. Stoiński, On superposition of the Bohr's almost periodic functions, *Funct. Approximation Comment. Math.* **5** (1977), 85–89.

[12] S. Stoiński, Real-valued functions almost periodic in variation, *Funct. Approximation Comment. Math.* **22** (1993), 141–148.

[13] S. Stoiński, A note on N-almost periodic functions and (NI)-almost periodic functions, *Commentat. Math.* **44** (2) (2004), 199–204.

In: Focus on Evolution Equations
Editor: Gaston M. N'Guerekata

ISBN: 978-1-60021-342-7

Chapter 11

RELATIVE STABILITY IN CONCAVE LAGRANGIAN SYSTEMS

Joël Blot[1,*]***, Biancamaria D'Onofrio***[2,†] ***and Roberto Violi***[3,‡]
[1]Laboratoire Marin Mersenne, Université Paris 1 Panthéon-Sorbonne, Paris, France
[2]Dipartimento di Matematica, Università di Roma La Sapienza, Roma, Italia
[3]Banca D'italia, Roma, Italia

Abstract

We study the asymptotic behaviour of solutions of a concave Lagrangian system, in presence of a discount rate, around a constant solution. Such systems are motivated by the macroeconomic theory. First by using an auxiliary function we exhibit invariant subsets and we obtain a result of unstability. Secondly we use the local theory of hyperbolic fixed points and consequences of the global concavity to obtain an existence and uniqueness result on the bounded solution which starts from a fixed state around the constant solution. We prove that these bounded solutions converge toward the constant solution, and moreover, under a sign condition, that these bounded solutions are optimal.

A.M.S. Classification: 34C11, 34D20, 34D45, 91B64, 37J25

Keywords: Euler-Lagrange equation, bounded solution, hyperbolic fixed point, stability, optimal growth theory.

Introduction

Let A be a nonempty subset of $\mathbf{R}^n \times \mathbf{R}^n$. Let $\ell : A \longrightarrow \mathbf{R}$ be a function. We assume that the following property holds

(1) ℓ is concave and of class C^3 on A.

We fix $\rho \in (0, \infty)$ and we consider the following Euler-Lagrange equation

*E-mail address: blot@univ-paris1.fr
†E-mail address: biancamariad@inwind.it
‡E-mail address: roberto.violi@bancaditalia.it

(EL) $\ell_x(x, \dot{x}) = \frac{d}{dt}\ell_{\dot{x}}(x, \dot{x}) - \rho.\ell_{\dot{x}}(x, \dot{x})$

where $\dot{x} := \frac{d}{dt}x$ and where ℓ_x and $\ell_{\dot{x}}$ denote the partial differentials of ℓ. This equation is associated to the following maximization problem

$$(P_\eta) \begin{cases} \text{Maximize} & J(x(.)) := \int_0^\infty e^{-\rho.t}\ell(x(t), \dot{x}(t))dt \\ \text{when} & x(.) \in \text{dom}J, x(0) = \eta \end{cases}$$

where η is fixed in $\mathbf{R}^n$ and where $\text{dom}J$ is the set of the functions $x(.) \in C^1(\mathbf{R}_+, \mathbf{R}^n)$ such that $(x(t), \dot{x}(t)) \in A$ for all $t \geq 0$ and such that the improper integral $\int_0^\infty e^{-\rho.t}\ell(x(t), \dot{x}(t))dt$ is convergent in $\mathbf{R}$, [6], [5]. This maximization problem is motivated by the Macroeconomic Optimal Growth Theory, [1], [2], [14]. The variational formalism for the Macroeconomic Optimal Growth Theory is developed in [13].

We shall assume that there exists $\overline{x} \in \mathbf{R}^n$, verifying $(\overline{x}, 0) \in A$, which is a constant solution of (EL), i.e.

(2) $\ell_x(\overline{x}, 0) = -\rho.\ell_{\dot{x}}(\overline{x}, 0)$

We study the asymptotic behaviour of the solution of (EL).

$BC^0(\mathbf{R}_+, \mathbf{R}^n)$ denotes the Banach space of the continuous bounded functions from $\mathbf{R}_+$ in $\mathbf{R}^n$, and $BC^1(\mathbf{R}_+, \mathbf{R}^n)$ denotes the Banach space of the functions $x(.) \in C^1(\mathbf{R}_+, \mathbf{R}^n)$ such that $x(.)$ and $\dot{x}(.)$ belong to $BC^0(\mathbf{R}_+, \mathbf{R}^n)$.

Definition. *A solution $x(.)$ of (EL) is called* an interior bounded solution of (EL) on $\mathbf{R}_+$ *when $x(.) \in BC^1(\mathbf{R}_+, \mathbf{R}^n)$ and when the closure of $\{(x(t), \dot{x}(t)) : t \in \mathbf{R}_+\}$ is included into A.*

Now we describe the contents of the paper. In Section 1 we use an auxiliary function to exhibit invariant subsets, to obtain a necessary condition to the boundedness of the solutions, and a result of unstability. In Section 2, under conditions which permit to use the stable manifold theorem, we provide condition to parametrize the stable manifold by the state variable. In Section 3 we establish a result of existence and of uniqueness of the bounded solution which starts from a given state around the constant solution. We prove that these bounded solutions converge toward the constant. Moreover under a sign condition, we prove that these bounded solutions are optimal.

1. An Auxiliary Function

We introduce two functions $V : A \longrightarrow \mathbf{R}$ and $G : A \longrightarrow \mathbf{R}$ defined by

(3) $V(x, \dot{x}) := \langle \ell_{\dot{x}}(x, \dot{x}) - \ell_{\dot{x}}(\overline{x}, 0), x - \overline{x} \rangle$

(4) $G(x, \dot{x}) := \langle D\ell(x, \dot{x}) - D\ell(\overline{x}, 0), (x, \dot{x}) - (\overline{x}, 0) \rangle$.

Under (2), $-D\ell$ is Minty-monotonic, [12], and consequently the inequality $G(x, \dot{x}) \leq 0$ holds everywhere. When $x(.)$ is a solution of (EL) on $\mathbf{R}_+$ we easily verify that the following relation holds

(5) $\frac{d}{dt}(e^{-\rho.t}V(x(t),\dot{x}(t))) = e^{-\rho.t}G(x(t),\dot{x}(t))$

And consequently we have proved the following result.

Lemma 1. *Under (1-2), when $x(.)$ is a solution of (EL) on $\mathbf{R}_+$, for all $t \geq 0$, we have*

(i) $G(x(t),\dot{x}(t)) \leq 0$

(ii) $t \longmapsto e^{-\rho.t}V(x(t),\dot{x}(t))$ *is non increasing on* $\mathbf{R}_+$.

Theorem 1. *Under (1-2) the sets* $[V \leq 0] := \{(x,\dot{x}) \in A : V(x,\dot{x}) \leq 0\}$ *and* $[V < 0] := \{(x,\dot{x}) \in A : V(x,\dot{x}) < 0\}$ *are invariant for the Lagrangian flow. Moreover when $x(.)$ is a solution of (EL) on $\mathbf{R}_+$, if there exists $t_0 \geq 0$ such that $V(x(t_0),\dot{x}(t_0)) < 0$ then $x(.)$ cannot be a bounded solution of (EL) on $\mathbf{R}_+$.*

Proof. Let $x(.)$ be a solution of (EL) on $\mathbf{R}_+$ and let $\tau \geq 0$ such that $V(x(\tau),\dot{x}(\tau)) \leq 0$ (respectively < 0). Then we have $e^{-\rho.\tau}V(x(\tau),\dot{x}(\tau)) \leq 0$ (respectively < 0), and by using Lemma 1 we obtain that, for all $t \geq \tau$, $e^{-\rho.t}V(x(t),\dot{x}(t)) \leq 0$ (respectively < 0), that implies $V(x(t),\dot{x}(t)) \leq 0$ (respectively < 0). And so the first assertion is proved. Denoting $v(t) := V(x(t),\dot{x}(t))$ and $g(t) := G(x(t),\dot{x}(t))$, a straightforward consequence of (5) is the following relation

(6) $\dot{v}(t) = \rho.v(t) + g(t).$

By using the variation of constants formula we obtain

(7) $v(t) = e^{\rho.(t-t_0)}.v(t_0) + e^{\rho.t}.\int_{t_0}^{t} e^{-\rho.s}.g(s)ds$

and consequently by using Lemma 1, for all $t \geq 0$, we have $v(t) \leq e^{\rho.(t-t_0)}.v(t_0)$. Since $v(t_0) < 0$ we obtain $v(t) \to -\infty$ when $t \to \infty$. But if $x(.)$ is a bounded solution of (EL) on $\mathbf{R}_+$, $\{V(x(t),\dot{x}(t)) : t \geq 0\}$ is bounded since a continuous function transforms compacts into compacts that forbids to have $v(t) \to -\infty$ when $t \to \infty$.■

The contraposition of the second assertion of Theorem 1 gives the following result which provides a necessary condition to the boundedness of solutions.

Corollary 1. *Under (1-2), if $x(.)$ is a bounded solution of (EL) on $\mathbf{R}_+$ then we have, for all $t \geq 0$,*

$$\langle \ell_{\dot{x}}(x(t),\dot{x}(t)) - \ell_{\dot{x}}(\overline{x},0), x(t) - \overline{x}\rangle \geq 0.$$

The following consequence of Theorem 1 is a result of unstability.

Corollary 2. *Under (1-2) and under the condition*

(8) $\ell_{\dot{x}\dot{x}}(\overline{x}, 0)$ *is negative definite,*

the constant solution $\overline{x}$ cannot be Liapunov stable and $\overline{x}$ cannot be an attractor for (EL).

Proof. We easily compute

(9) $V(\overline{x}, 0) = 0$ and $DV(\overline{x}, 0) = 0$.

Moreover we obtain

$$D^2V(\overline{x}, 0)((h, k), (h, k)) = 2\ell_{\dot{x}\dot{x}}(\overline{x}, 0).(h, k) + 2\ell_{\dot{x}x}(\overline{x}, 0).(h, h).$$

Consequently when we fix $h \neq 0$, by taking $k = \alpha h$ with $\alpha > 0$, α big enough, we obtain $D^2V(\overline{x}, 0)((h, k), (h, k)) < 0$, and by taking $k = \beta h$, with $\beta < 0$, $|\beta|$ big enough, we obtain $D^2V(\overline{x}, 0)((h, k), (h, k)) > 0$. And so we have proved

(10) $D^2V(\overline{x}, 0)$ takes positive values and negative values.

If $(h, k) \in \text{Ker}D^2V(\overline{x}, 0)$ then we have $2\ell_{\dot{x}x}(\overline{x}, 0).h + \ell_{\dot{x}\dot{x}}(\overline{x}, 0).k = 0$ and $\ell_{\dot{x}\dot{x}}(\overline{x}, 0).h = 0$. By using (8) we obtain $h = 0$ by the second equation, and $k = 0$ by the first equation. And so we have

(11) $D^2V(\overline{x}, 0)$ is invertible.

By using (9) and (11) we can apply the Morse Lemma at V, [11]. And consequently there exist a neighborhood ω of $(\overline{x}, 0)$ in $\mathbf{R}^n \times \mathbf{R}^n$, a neighborhood ω_1 of $(0, 0)$ in $\mathbf{R}^n \times \mathbf{R}^n$, and a diffeomorphism $\phi : \omega \longrightarrow \omega_1$ such that, for all $(x, \dot{x}) \in \omega$,

$$V(x, \dot{x}) = \frac{1}{2} D^2V(\overline{x}, 0).(\phi(x, \dot{x}), \phi(x, \dot{x})).$$

Then by using (10) we can say

(12) On each neighborhood of $(\overline{x}, 0)$ in $\mathbf{R}^n \times \mathbf{R}^n$, V takes positive values and negative values.

If $\overline{x}$ is Liapunov stable or if it is an attractor for (EL), there exists a neighborhood ω_0 of $(\overline{x}, 0)$ in $\mathbf{R}^n \times \mathbf{R}^n$ such that, for each $(x_0, \dot{x}_0) \in \omega_0$ and for each solution $x(.)$ of (EL) verifying $x(0) = x_0$ and $\dot{x}(0) = \dot{x}_0$, the distance between $(x(t), \dot{x}(t))$ and $(\overline{x}, 0)$ remains bounded on $\mathbf{R}_+$. Consequently $x(.)$ remains bounded on $\mathbf{R}_+$, and by using Corollary 1 we have $V(x_0, \dot{x}_0) \geq 0$. And so we obtain $V \geq 0$ on ω_0 which is impossible after (12).■

The conclusion of Corollary 2 is more or less a known result when the Lagrangian system is transformable into a Hamiltonian system with a concave-convex Hamiltonian function, [7] Chapters 3, 4, 6. Here we do not need additional assumptions to transform the Lagrangian system in a Hamiltonian system.

Theorem 1 treats the case where $t \mapsto V(x(t), \dot{x}(t))$ takes negative values. The complementary case is this one where all the values of $V(x(t), \dot{x}(t))$ are non negative. By using Lemma 1 and the monotonic limit theorem we see that the limit $\lim_{t\to\infty} e^{-\rho t} V(x(t), \dot{x}(t))$ exits ever in $\mathbf{R}_+$, and by using (7) we see that this limit is equal to $V(x(0), \dot{x}(0)) + \int_0^\infty e^{-\rho t} G(x(t), \dot{x}(t))dt$. And so we distinguish two subcases: this limit is positive and this limit is equal to zero.

Proposition 1. *Under (1-2) let $x(.)$ be a solution of (EL) on $\mathbf{R}_+$ such that $\lim_{t\to\infty} e^{-\rho t} V(x(t), \dot{x}(t)) > 0$, i.e. $V(x(0), \dot{x}(0)) > -\int_0^\infty e^{-\rho t} G(x(t), \dot{x}(t))dt$, then $x(.)$ cannot be an interior bounded solution.*

Proof. The assumption implies $\lim_{t\to\infty} V(x(t), \dot{x}(t)) = \infty$. We proceed by contradiction, we assume that $x(.)$ is an interior bounded solution, then, since V is continuous and since the closure of $\{(x(t), \dot{x}(t)) : t \in \mathbf{R}_+\}$ is compact, $V(x(t), \dot{x}(t))$ is bounded on $\mathbf{R}_+$ and consequently it cannot converge to ∞ when t converges to ∞.■

In the second subcase, $\lim_{t\to\infty} e^{-\rho t} V(x(t), \dot{x}(t)) = 0$, i.e. $V(x(0), \dot{x}(0)) = -\int_0^\infty e^{-\rho t} G(x(t), \dot{x}(t))dt$, even if in addition we assume that $\lim_{t\to\infty} G(x(t), \dot{x}(t)) = g_*$ is finite, then we have, for all $t \geq 0$,

$$V(x(t), \dot{x}(t)) = \frac{V(x(0), \dot{x}(0)) + \int_0^t e^{-\rho t} G(x(t), \dot{x}(t))dt}{e^{-\rho t}},$$

and

$$\frac{\frac{d}{dt}\left[V(x(0), \dot{x}(0)) + \int_0^t e^{-\rho t} G(x(t), \dot{x}(t))dt\right]}{\frac{d}{dt}\left[e^{-\rho t}\right]} = \frac{e^{-\rho t} G(x(t), \dot{x}(t))}{-\rho e^{-\rho t}} = \frac{G(x(t), \dot{x}(t))}{-\rho}.$$

and by using the l'Hospital rule we obtain $\lim_{t\to\infty} V(x(t), \dot{x}(t)) = \frac{-1}{\rho} g_*$ which is finite. But in general it is not clear that we are able to deduce of this finitness some consequences on the asymptotic behaviour of $x(t)$ without special conditions on the function $\ell_{\dot{x}}$.

2. A Local Study

Under the assumption (8) by using a classical Hilbert theorem of Calculus of Variations [9], we know that (EL) becomes the following second-order ordinary differential equation

(13) $\ddot{x} = (\ell_{\dot{x}\dot{x}}(x, \dot{x}))^{-1}(\ell_x(x, \dot{x}) + \rho.\ell_{\dot{x}}(x, \dot{x}) - \ell_{\dot{x}x}(x, \dot{x}).\dot{x})$.

Introducing the vector field $F : A \longrightarrow \mathbf{R}^n \times \mathbf{R}^n$

$$F(x, y) := (y, (\ell_{\dot{x}\dot{x}}(x, y))^{-1}(\ell_x(x, y) + \rho.\ell_{\dot{x}}(x, y) - \ell_{\dot{x}x}(x, y).y))$$

the equation (13) becomes the first-order differential equation

(14) $(\dot{x}, \dot{y}) = F(x, y)$.

We formulate the following additional assumption

(15) $DF(\overline{x}, 0)$ possesses n eigenvalues (distinct or not) with a negative real part and it possesses n eigenvalues (distinct or not) with a positive real part.

Under this assumption, $(\overline{x}, 0)$ is a hyperbolic constant solution of (14) [10]. This permits to use the Hartman-Grobman theorem and the Stable Manifold Theorem [10]. A such approach is used in [8] to exhibit periodic solutions.

In the more classical macroeconomic example [1] where $n = 1$, $\ell(x, \dot{x}) := \frac{1}{\sigma}(x^\alpha - \dot{x})^\sigma$ with $0 < \alpha < 1$, $0 < \sigma < 1$, $x > 0$, $x^\alpha - \dot{x} > 0$, we have

$$DF(\overline{x}, 0) = \begin{bmatrix} 0 & 1 \\ \frac{\alpha(\alpha-1)(\overline{x})^{2\alpha-2}}{\sigma-1} & \rho \end{bmatrix}$$

Consequently $\det DF(\overline{x}, 0) = -\frac{\alpha(\alpha-1)(\overline{x})^{2\alpha-2}}{\sigma-1} < 0$, and so if λ_1 and λ_2 denote the eigenvalues of $DF(\overline{x}, 0)$, we have $\lambda_1.\lambda_2 < 0$ that implies that λ_1 and λ_2 are real and they have different signs. And so (15) is fulfilled.
Furthermore for a general one-sector optimal growth model, with representative utility function u with standard properties $u' \geq 0$ and $u'' < 0$, and a production ψ with $\psi' \geq 0$ and $\psi'' < 0$, [2] p. 38-47, the Lagrangian function is $\ell(x, \dot{x}) := u(\psi(x) - \dot{x})$. Around a constant solution $(\overline{x}, 0)$ of (EL), in this case we obtain

$$DF(\overline{x}, 0) = \begin{bmatrix} 0 & 1 \\ \psi''(\overline{x}).\frac{u'(\psi(\overline{x}))}{u''(\psi(\overline{x}))} & \rho \end{bmatrix}$$

and $\det DF(\overline{x}, 0) = -\psi''(\overline{x}).\frac{u'(\psi(\overline{x}))}{u''(\psi(\overline{x}))} < 0$. And condition (15) is also fulfilled.

Notations. ([10]) E^s denotes the vector subspace of $\mathbf{R}^n \times \mathbf{R}^n$ generated by the spectral subspaces associated to the eigenvalues of $DF(\overline{x}, 0)$ which possess a negative real part. W^s_{loc} is the local n-dimensional submanifold provided by the Stable Manifold Theorem. If $(\varphi_t)_t$ denotes the flow of (14), U being a neighborhood of $(\overline{x}, 0)$ in $\mathbf{R}^n \times \mathbf{R}^n$, we have

$$W^s_{\text{loc}} = \{(x, y) \in U : \lim_{t \to \infty} \varphi_t(x, y) = (\overline{x}, 0)\}.$$

We formulate the following additional assumption on E^s

(16) $(0 \times \mathbf{R}^n) \cap E^s = \{(0, 0)\}$.

Note that this condition is fulfilled in the previous economic example. This condition will permit to obtain a parametrization of W^s_{loc} by $\mathbf{R}^n \times 0$ as we shall see it in the following lemma.

Lemma 2. *Under (1), (2), (8), (15), (16), there exist a neighborhood A_0 de ($\overline{x}$ in $\mathbf{R}^n$, a neighborhood of 0 in $\mathbf{R}^n$, and there exists a function $\zeta \in C^1(A_0, B)$ such that*

$$(A_0 \times B) \cap W^s_{\text{loc}} = \{(x, \zeta(x)) : x \in A_0\}.$$

Proof. Let $\pi_1(x, \dot{x}) := x$ be the projection on the first factor from $\mathbf{R}^n \times \mathbf{R}^n$ in $\mathbf{R}^n$. Then we have $\varpi_1 := \pi_{1|W^s_{\text{loc}}} \in C^1(W^s_{\text{loc}}, \mathbf{R}^n)$. We know that the tangent space to the manifold W^s_{loc} at $(\overline{x}, 0)$ is equal to E^s, [10]. And so the differential $D\varpi_1(\overline{x}, 0) \in \mathcal{L}(E^s, \mathbf{R}^n)$ and we have $D\varpi_1(\overline{x}, 0) = D\pi_1(\overline{x}, 0)_{|E^s} = \pi_{1|E^s}$. Since $\pi_{1|E^s}$ is a linear mapping between vector spaces having the same finite dimension, it is bijective if and only if it is injective. We note that $\text{Ker}\pi_{1|E^s} = E^s \cap (0 \times \mathbf{R}^n) = \{(0,0)\}$ following (16), and so $\pi_{1|E^s}$ is injective, and consequently it is bijective. Then by using the inverse function theorem, ϖ_1 is a local C^1-diffeomorphism, and $\varpi_1{}^{-1}$ necessarily is in the form $x \longmapsto (x, \zeta(x))$ that ensures the conclusion.■

In [8] the author use additional assumptions, notably on the multiplicity of the eigenvalues of $DF(\overline{x}, 0)$ and he does not condition as our condition (16) to ensure a good parametrization of the stable manifold.

3. Attractivity and Optimality

The following lemma is a straightforward consequence of Theorem 1 of [3].

Lemma 3. *Under (1), (2) and (8), for all $x \in \mathbf{R}^n$, there is at most one solution of (EL) which is bounded on $\mathbf{R}_+$ and which takes the value x at the time $t = 0$.*

Theorem 2. *We assume the conditions (1), (2), (8), (15) and (16) fulfilled.*

- **(i)** *There exists a neighborhood A_0 of $\overline{x}$ in $\mathbf{R}^n$ such that, for all $x \in A_0$, there exists a unique solution $x(.)$ of (EL) which is bounded on $\mathbf{R}_+$ and which satisfies $x(0) = x$. Moreover we have $\lim_{t\to\infty} x(t) = \overline{x}$ and $\lim_{t\to\infty} \dot{x}(t) = 0$.*

- **(ii)** *If in addition we assume that $\ell \geq 0$ then, for all $x \in A_0$, the unique solution $x(.)$ of (EL) which is bounded on $\mathbf{R}_+$ and which satisfies $x(0) = x$ is an optimal solution of (P_x).*

Proof. We consider the neighborhood A_0 provided by Lemma 2 and we arbitrarily fix $x \in A_0$. Then the solution $t \longmapsto \varphi_t(x, \zeta(x))$ is bounded on $\mathbf{R}_+$ since $\lim_{t\to\infty} \varphi_t(x, \zeta(x)) = (\overline{x}, 0)$. By using Lemma 3, it is the only solution of a such type. That proves (i).
The assertion (ii) is a consequence of (i) and of Theorem 1 of [4].■

To conclude we give some comments on the existing results about these questions. These existing results are formulated in the hamiltonian formalism. The local result of Brock and Sheinkman, [7] p. 132, uses the value-fonction of the maximization problem and need the differentiability of classe C^2 of the value-function. The two other following results are global: this one of Cass and Shell, [7] p. 128, consider only solutions which satisfy the condition $\lim_{t\to\infty} e^{-\rho t}\langle \ell_{\dot{x}}(x(t), \dot{x}(t)), x(t)\rangle = 0$, and the global result of Brock and Sheinkman, [7] p. 129, uses a strengthened condition of concavity.

References

[1] R.T. Barro and X. Sala-I-Martin, "La croissance économique", French Edition, Ediscience International and McGrawhill Book Co, Europe, 1996.

[2] O.J. Blanchard and S. Fischer, "Lectures on Macroeconomics", M.I.T. Press, Cambridge, Massachusetts, 1989.

[3] J. Blot and P. Cartigny, Bounded Solutions and Oscillations of Convex Lagrangian Systems in Presence of a Discount Rate, *Zeitschrift für Analysis und ihre Anwendungen*, **14** (1995), 731-750.

[4] J. Blot and P. Cartigny, Optimality in Infinite-Horizon Problems under Signs Conditions, *Journal of Optimization Theory and Applications*, **106** (2000), 411-419.

[5] J. Blot and N. Hayek, Second-Order Necessary Conditions for the Infinite-Horizon Variational Problems, *Mathematics of Operations Research*, **21** (1996), 979-990.

[6] J. Blot and P. Michel, First-Order Necessary Conditions for the Infinite-Horizon Variational Problems, *Journal of Optimization Theory and Applications*, **88** (1996), 339-364.

[7] D.A. Carlson, A.B. Haurie, A. Leizarowitz, "Infinite Horizon Optimal Control", Second, Revised and Enlarged Edition, Springer-Verlag, Berlin, 1991.

[8] P. Cartigny, Stabilité des solutions stationnaires et oscillations dans un problème variationnel : approche lagrangienne, *C. R. Acad. Sci. Paris*, **318** (1994), 869-872.

[9] G.E. Ewing, "Calculus of Variations and Applications", Dover Publ., Inc., New York, 1985.

[10] J. Guckenheimer and P. Holmes, "Nonlinear Oscillations, Dynamical Systems, and Bifurcations of Vector Fields", Springer-Verlag, New York, 1983.

[11] L. Nirenberg, "Topics in Nonlinear Functional Analysis", *Courant Lecture Notes in Mathematics,* A.M.S., Providence, Rhode Island, 2001.

[12] R.T. Rockafellar, "Convex Analysis", Princeton Univ. Press, Princeton, N. J., 1970.

[13] T.S. Sargent, "Macroeconomic Theory", Second Edition, Academic Press, Inc., Cambridge, MA, 1987.

In: Focus on Evolution Equations
Editor: Gaston M. N'Guerekata

ISBN: 978-1-60021-342-7

Chapter 12

QUASISEMIGROUPS AND EVOLUTION EQUATIONS*

Diómedes Bárcenas[1,†,‡], ***Hugo Leiva***[1,§] ***and Ambrosio Tineo Moya***[2,¶]
[1]Departamento de Matemáticas, Universidad de Los Andes,
Mérida 5101, Venezuela
[2]Departamento de Física y Matemáticas,
Núcleo Universitario Rafael Rangel, Universidad de los Andes,
Trujillo 3102, Venezuela

Abstract

We study the notion of Quasisemigroup of bounded linear operators as a generalization of that of strongly continuous semigroup and give some applications in abstract evolution equations.

1991 Mathematics Subject Classification: Primary 54C40, 14E20; Secondary 46E25, 20C20

Keywords and phrases: Quasisemigroup, generator.

1. Introduction

The theory of Quasisemigroups of bounded linear operators appeared the first time in (1991),([2]), a preprint written by the first two authors; and the idea to write a paper in a definitive form came from the facts that its study has been undertaken by Cuc in several papers ([3],[4],[5]). Given a Banach space X, by $L(X)$ we denote the space of all bounded linear operators from X to X. A bi-parametric family of bounded linear operators $\{K(t,s)\}_{t,s\geq 0} \subset L(X)$ is called commutative if it satisfies:

$$K(r,t+s) = K(r+t,s)K(r,t) = K(r,t)K(r+t,s).$$

---

*This paper is dedicated to our authors

†E-mail address: barcenas@ula.ve

‡Supported by CDCHT of Universidad de Los Andes under projects C-291-01-05-D.

§E-mail address: hleiva@ula.ve

¶E-mail address: atemoya@ula.ve o ambrosiotineo@yahoo.com

A commutative family $\{K(t,s)\}_{t,s\geq 0}$ is called strongly continuous quasisemigroup if it satisfies:

i) $K(t,0) = I \quad (t \geq 0)$ (I is the identity operator in $L(X)$).

ii) $K(r,t+s) = K(r+t,s)K(r,t) \quad (r,s,t \geq 0)$.

iii) $\lim\limits_{(t,s)\to(t_0,s_0)} \|K(t,s)x_0 - K(t_0,s_0)x_0\| = 0 \qquad (x_0 \in X)$.

iv) There exists a continuous and increasing function $M : [0,\infty) \longrightarrow [1,\infty)$ such that

$$\|K(t,s)\| \leq M(t+s) \quad \text{for every} \quad t,s \geq 0.$$

Also, we introduce the notion of generator $A(t)$ of a strongly continuous quasisemigroup. This two notions generalize the notions of strongly continuous semigroup and its infinitesimal generator.

Given a commutative evolution operator $U(t,s)$, we associate to it generates a strongly continuous quasisemigroup defined by

$$K(t,s) = U(t+s,t),$$

which frequently appear in the study of evolution equations.

We prove the following statements:
If $A(t)$ with domain D generates a strongly continuous quasisemigroup $K(t,s)$ on a Banach space X (see definition 2.2), then for each $x_0 \in D$ and $r \geq 0$, the evolution equation with initial conditions

$$\begin{cases} \dot{x}(t) = A(r+t)x(t) \\ x(0) = x_0, \end{cases}$$

admits the function $x(t) = K(t,r)x_0$ as a unique solution. Furthermore, the system

$$\begin{cases} \dot{x}(t) = A(r+t)x(t) + f(t), & (0 \leq t \leq T) \\ x(0) = x_0, & (x_0 \in D), \end{cases}$$

where $A(t)$ is strongly continuous and $f(\cdot)$ is continuously differentiable on $[0,T]$ with values in D admits only one mild solution given by

$$x(t) = K(r,t)x_0 + \int_0^t K(r+s,t-s)f(s)ds.$$

Finally, if $A(t)$ is the generator of a strongly continuous quasisemigroup $K(t,s)$ on a Banach space X and $B \in L(X)$, then $A(t)+B$ is the generator of the strongly continuous quasisemigroup $R(t,s)$ defined by

$$R(r,t) = K(r,t)x_0 + \int_0^t K(r+s,t-s)BR(r,s)x_0 ds, \quad x_0 \in X.$$

2. Quasisemigroup of Operators

Definition 2.1 Let X be a Banach space and $L(X)$ the space of linear and bounded operators from X to X. A family of operators $K(t,s) \in L(X)$, $t, s \geq 0$, is called a **Strongly Continuous Quasisemigroup** in X if it is commutative and verifies:

1) $K(t,0) = I \quad (t \geq 0) \quad$ (I is the identity operator in $L(X)$).

2) $K(r, t+s) = K(r+t, s)K(r,t) \quad (t, r, s \geq 0)$.

3) $\lim_{(t,s)\to(t_0,s_0)} \|K(t,s)x_0 - K(t_0,s_0)x_0\| = 0 \qquad (x_0 \in X)$.

4) There exists a continuous and increasing function $M : [0,\infty) \longrightarrow [1,\infty)$ such that

$$\|K(t,s)\| \leq M(t+s) \quad \text{for every} \quad t, s \geq 0.$$

Definition 2.2 Let $K(t,s)$ be a strongly continuous quasisemigroup and suppose that D is a dense subspace of X, such that for all $x_0 \in D$ there exist the limits:

$$\lim_{s\to 0^+} \frac{K(t,s)x_0 - x_0}{s} = \lim_{s\to 0^+} \frac{K(t-s,s)x_0 - x_0}{s}, \qquad t > 0,$$

$$\lim_{s\to 0} \frac{K(0,s)x_0 - x_0}{s}.$$

The family of operators $A(t)$, $t \geq 0$, with common domain D, defined by

$$A(t)x_0 = \lim_{s\to 0^+} \frac{K(t,s)x_0 - x_0}{s} \qquad (x_0 \in D),$$

is called the **Generator of the Quasisemigroup** $K(t,s)$.

Now, we give some example of quasisemigroups.

Example 2.1 Let $\{T_t\}$, $t > 0$, be a strongly continuous semigroup on a Banach space X and let A be the infinitesimal generator . If $K(t,s) = T_s$ for all $t, s \geq 0$, then $K(t,s)$ is a strongly continuous quasisemigroup whose generator is $A(t) = A$, and $D = D(A)$ which is a dense subspace in X.

Example 2.2 If we denote by X the Banach space of the uniformly continuous and bounded real functions defined on $[0,\infty)$, with the supremum norm, the family of operators $K(t,s) \in L(X)$ defined by

$$(K(t,s)x)(\xi) = x(s^2 + 2st + \xi) \quad (t, s \geq 0)$$

is a strongly continuous quasisemigroup of contraction, where the generator is given by $A(t)x = 2t\dot{x}$, with $x \in D = \{x \in X : \dot{x} \in X\}$, which is a dense subspace in X.

Example 2.3 Let $\{T_t\}$, $t \geq 0$, be a strongly continuous semigroup on a Banach space X and A its infinitesimal generator. The family of linear and bounded operators

$$K(t,s) = e^{(T_{t+s}-T_t)} \quad (t, s \geq 0),$$

is a strongly continuous quasisemigroup with generator $A(t) = AT_t$ and $D = D(A)$ which is a dense subspace in X.

Example 2.4 Let $\{T_t\}$, $t \geq 0$, be a strongly continuous semigroup on a Banach space X and A its infinitesimal generator. The family of linear and bounded operators

$$K(t,s) = T(g(t+s) - g(t)) \quad (t, s \geq 0),$$

where $g(t) = \int_0^t a(s)ds$ and $a(t) > 0$ a continuous function on $[0, \infty)$, is a strongly continuous quasisemigroup with generator $A(t) = a(t)A$ and $D = D(A)$ which is a dense subspace in X.

Remark 2.1 The **Evolution Operator** associated with

$$\begin{cases} \dot{x}(t) = A(t)x(t), & (t > 0) \\ x(0) = x_0, & (x_0 \in X) \end{cases} \tag{2.1}$$

is defined in [6], [11], and [13] as the family $U(t,s) \in L(X)$, $0 \leq s \leq t < \infty$, with the following properties

1) $U(r,r) = I$, (I is the identity operator on $L(X)$).
2) $U(t,r)U(r,s) = U(t,s)$, $(0 \leq s \leq r \leq t < \infty)$.
3) $U(\cdot,\cdot)$ is strongly continuous.

A commutative evolution operator $U(t,s)$ satisfying the condition

$$\|U(t,s)\| \leq M(t+s)$$

induces the quasisemigroup given by

$$K(t,s) = U(t+s,t) \qquad (0 \leq t \leq t+s < \infty).$$

Indeed,
1) $K(t,0)x_0 = U(t,t)x_0 \implies K(t,0) = I$, in $L(X)$.

2)

$$\begin{aligned} K(t, r+s)x_0 &= U(t+r+s,t)x_0 = U(t+r+s,t+r)U(t+r,t)x_0 \\ &= K(t+r,s)K(t,r)x_0. \end{aligned}$$

3) Since

$$K(t,s)x_0 - K(t_0,s_0)x_0 = U(t+s,t)x_0 - U(t_0+s_0,t_0)x_0,$$

we see that

$$\lim_{(t,s)\to(t_0,s_0)} K(t,s)x_0 = K(t_0,s_0)x_0.$$

The following identities will be used frequently in this work

$$\begin{aligned} K(r,t+s) &= K(r+t,s)K(r,t) \\ K(r,t) &= K(r+t-s,s)K(r,t-s), \quad (t>s). \end{aligned} \tag{2.2}$$

Theorem 2.1 *Let $A(t)$ the generator of a strongly continuous quasisemigroup $K(t,s)$ and D the domain of $A(t)$. Then*
a) If $x_0 \in D$, then $K(t,s)x_0 \in D$ for every $t, r \geq 0$.

b) For each $x_0 \in D$ and $r \geq 0$,

$$\frac{\partial K(r,t)x_0}{\partial t} = A(r+t)K(r,t)x_0 = K(r,t)A(r+t)x_0.$$

c) If $A(\cdot)$ is local strongly integrable, then for each $x_0 \in D$ and $r \geq 0$, we have

$$K(r,t)x_0 = x_0 + \int_0^t A(r+s)K(r,s)x_0 ds, \quad (t \geq 0).$$

Proof. 2.1 Let $r, t \geq 0$ be fixed and $s > 0$. Since $x_0 \in D$, then

$$\lim_{s\to 0^+} \frac{K(t,s)x_0 - x_0}{s}$$

exists and it is equals $A(t)x_0$. For $s > 0$, let us consider

$$\frac{K(t,s)K(r,t)x_0 - K(r,t)x_0}{s} = (\frac{K(t,s)-I}{s})K(r,t)x_0 = K(r,t)(\frac{K(t,s)-I}{s})x_0.$$

Since the quasisemigroup is commutative and $x_0 \in D$, then the limits exist as $s \to 0^+$. Therefore, $K(r,t)x_0 \in D$ and $A(t)K(r,t)x_0 = K(r,t)A(t)x_0$.
To prove the part b), it is enough to prove the existence and coincidence of both right and left derivative. Let $x_0 \in D$ and $t > r \geq 0$.
Consider the quotient

$$\begin{aligned} \frac{K(r,t+s)x_0 - K(r,t)x_0}{s} &= \frac{K(r+t,s)K(r,t)x_0 - K(r,t)x_0}{s} \\ &= K(r,t)\Big(\frac{K(r+t,s)x_0 - x_0}{s}\Big). \end{aligned}$$

Taking limit as $s \to 0^+$, the partial derivative exists and it is equal to $K(r,t)A(r+t)x_0$.
For $s < 0$,

$$\begin{aligned} \frac{K(r,t+s)x_0 - K(r,t)x_0}{s} &= \frac{K(r,t)x_0 - K(r,t+s)x_0}{-s} \\ &= \frac{K(r+t+s,-s)K(r,t+s)x_0 - K(r,t+s)x_0}{-s} \\ &= \frac{K(r,t+s)\Big(K(r+t+s,-s)x_0 - x_0\Big)}{-s}. \end{aligned}$$

Passing to the limit as $s \to 0^-$ $(-s \to 0^+)$ and using the strong continuity of quasisemigroup, the limit exists and is equal to $K(r,t)A(r+t)x_0$, which proves part b).
To prove part c), let be $x_0 \in D$ and $t > 0$. We define the vectorial function by

$$g(t) = K(r,t)x_0.$$

The function g is differentiable and from b) we obtain

$$\frac{dg(t)}{dt} = \frac{\partial K(r,t)x_0}{\partial t} = A(r+t)K(r,t)x_0 = K(r,t)A(r+t)x_0.$$

Therefore, applying the fundamental theorem of calculus we get

$$\int_s^t K(r,u)A(r+u)x_0 du = g(t) - g(s) = K(r,t)x_0 - K(r,s)x_0.$$

Putting $s = 0$ in the previous equation we obtains the result. Thus

$$K(r,t)x_0 = x_0 + \int_0^t A(r+u)K(r,u)x_0 du.$$

Theorem 2.2 *Let $A(t)$ be the generator of a strongly continuous quasisemigroup $K(t,s)$ on a Banach space X and D the domain of $A(t)$. Then for each $x_0 \in D$ and $r \geq 0$, the initial value problem:*

$$\begin{cases} \dot{x}(t) = A(r+t)x(t) \\ x(0) = x_0 \end{cases} \tag{2.3}$$

admits a unique solution.

Proof. 2.2 Let us define the following function

$$x(t) = K(r,t)x_0.$$

By Theorem 2.1, the function $x(t)$ is solution of the initial value problem (2.3), indeed,

$$\dot{x}(t) = \frac{\partial K(r,t)x_0}{\partial t} = A(r+t)K(r,t)x_0 = K(r,t)A(r+t)x_0.$$

Let us suppose that $y(t)$ is another solution of the initial value problem (2.3) and let us consider the function

$$F(s) = K(r+s, t-s)y(s) \quad s \in [0,t].$$

Consider the following function

$$G(s) = K(r+t, t-s)x_0.$$

Now

$$\frac{G(s+h) - G(s)}{h} = \frac{K(r+s+h, t-s-h)x_0 - K(r+s, t-s)x_0}{h}. \tag{2.4}$$

Using the identity (2.2), we obtain

$$\begin{aligned} K(r+s,t-s) &= K(r+s+t-s-(t-s-h),t-s-h)K(r+s,t-s-(t-s-h)) \\ &= K(r+s+h,t-s-h)K(r+s,h) \qquad (2.5) \end{aligned}$$

Substituting (2.5) in the equation (2.4), we get

$$\begin{aligned} \frac{G(s+h)-G(s)}{h} &= \frac{K(r+s+h,t-s-h)x_0 - K(r+s+h,t-s-h)K(r+s,h)x_0}{h} \\ &= -K(r+s+h,t-s-h)\Big[\frac{K(r+s,h)x_0 - x_0}{h}\Big] \end{aligned}$$

Passing to the limit as $h \to 0$, we obtain

$$\dot{G}(s) = -K(r+s,t-s)A(r+s)x_0,$$

since

$$F(s) = G(s)y(s),$$

then

$$\begin{aligned} \dot{F}(s) &= -K(r+s,t-s)A(r+s)y(s) + K(r+s,t-s)\dot{y}(s) \\ &= -K(r+s,t-s)A(r+s)y(s) + K(r+s,t-s)A(r+s)y(s) \\ &= 0. \end{aligned}$$

In this way, we have that $\dot{F}(s) = 0$ for each $s \in (0,t)$. Consequently, $F(s)$ is a constant function for every $s \in [0,t]$. In particular, $F(0) = F(t)$. Since $F(0) = K(r,t)y(0) = K(r,t)x_0$ and $F(t) = K(r+t,0)y(t) = y(t)$, we conclude that $y(t) = K(r,t)x_0$, which proves the uniqueness of the solution. Let us observe that

$$\frac{\partial K(r+s,t-s)}{\partial s} = -K(r+s,t-s)A(r+t).$$

Proposition 2.1 *Let $K(s,t)$ be a strongly continuous quasisemigroup on the Banach space X. If $f : [0,T] \longrightarrow X$ is continuous function, then*

$$\lim_{h\to 0}\frac{1}{h}\int_t^{t+h} f(u)K(s,u)x_0 du = f(t)K(s,t)x_0.$$

Proof. 2.3 Consider $s,t \geq 0$, and $x_0 \in X$. We define the following function

$$\varphi(t) = f(t)K(s,t)x_0.$$

φ is clearly continuous in t. Hence

$$F(\epsilon) = \int_t^{t+\epsilon} \varphi(u)du \quad (\epsilon > 0)$$

is well defined. Now,

$$F'(0) = \lim_{h\to 0} \frac{F(0+h) - F(0)}{h} = \lim_{h\to 0} \frac{1}{h}\int_t^{t+h} \varphi(u)du, \quad (h > 0)$$

and for each $\epsilon > 0$ we have

$$F'(\epsilon) = \frac{d}{d\epsilon}\int_t^{t+\epsilon} \varphi(u)du = \varphi(t+\epsilon) = f(t+\epsilon)K(s,t+\epsilon)x_0.$$

Putting $\epsilon = 0$,

$$F'(0) = f(t)K(s,t)x_0.$$

Therefore,

$$\lim_{h\to 0} \frac{1}{h}\int_t^{t+h} \varphi(u)du = f(t)K(s,t)x_0.$$

Proposition 2.2 *Let $A(t)$ with domain D be the generator of a strongly continuous quasisemigroup $K(s,t)$ and $f : [0,T] \longrightarrow X$ a continuous function. If $\{x_n\}$ is a sequence in D such that $x_n \to x$ and $A(t)x_n$ converges uniformly to $f(t)$ in $[0,T]$, then for each $0 \leq r < T$, we have that $A(r)x = f(r)$.*

Proof. 2.4 Let be $r \in [0,t]$, and $s > 0$. By definition,

$$A(r)x_n = \lim_{s\to 0^+} \frac{K(r,s)x_n - x_n}{s}.$$

Passing to the limit as $n \to \infty$, $A(r)x_n \to A(r)x$. Now, from the uniform convergence and the Theorem 2.1,

$$\begin{aligned} K(r,s)x - x &= \lim_{n\to\infty} \Big(K(r,s)x_n - x_n\Big) \\ &= \lim_{n\to\infty} \int_0^s K(r,u)A(r+u)x_n du \\ &= \int_0^s K(r,u)f(r+u)du. \end{aligned}$$

Dividing by $s > 0$,

$$\frac{K(r,s)x - x}{s} = \frac{1}{s}\int_0^s K(r,u)f(r+u)du,$$

and taking limits as $s \to 0^+$

$$\begin{aligned} A(r)x = \lim_{s\to 0^+} \frac{K(r,s)x - x}{s} &= \lim_{s\to 0^+} \frac{1}{s}\int_0^s K(r,u)f(r+u)du \\ &= K(r,0)f(r+0) = f(r). \end{aligned}$$

Theorem 2.3 *Let $K(s,t)$ be a strongly continuous quasisemigroup on a Banach space X with $A(t)$ strongly continuous. If $f:[0,T]\longrightarrow D$ is a continuously differentiable function and*

$$\int_0^t K(r+s,t-s)f(s)ds \in D, \qquad (0\le t\le T).$$

Then, the initial value problem

$$\begin{cases} \dot{x}(t) = A(r+t)x(t) + f(t), & (0<t\le T) \\ x(0) = x_0, & (x_0\in D) \end{cases} \tag{2.6}$$

admits as unique solution, the function

$$x(t) = K(r,t)x_0 + \int_0^t K(r+s,t-s)f(s)ds. \tag{2.7}$$

Proof. 2.5 For the existence of the solution it is enough to prove that the function $x(t)$ in (2.7) has strong derivative and satisfies the equation (2.6).
Clearly $x(0)=x_0$. On the other hand, the function $y(t)=K(r,t)x_0$ is a solution of the problem of value initial (2.3). So we need prove that

$$\int_0^t K(r+s,t-s)f(s)ds$$

satisfies the initial value problem (2.6). Indeed, let us define the function

$$g(t) = \int_0^t K(r+s,t-s)f(s)ds. \tag{2.8}$$

Changing variables, $g(t)$ can be written as follows

$$g(t) = \int_0^t K(r+s,s)f(t-s)ds. \tag{2.9}$$

Since $K(s,t)$ is a strongly continuous quasisemigroup and $f(\cdot)$ is continuous, the above integral exists. Let $h\neq 0$ and consider the following quotient

$$\begin{aligned} \frac{g(t+h)-g(t)}{h} &= \frac{1}{h}\Big[\int_0^{t+h} K(r+s,s)f(t+h-s)ds - \int_0^t K(r+s,s)f(t-s)ds\Big] \\ &= \frac{1}{h}\Big[\int_0^t K(r+s,s)\Big(f(t+h-s)-f(t-s)\Big)ds \\ &+ \int_t^{t+h} K(r+s,s)f(t+h-s)ds\Big]. \end{aligned}$$

Taking limits when $h\to 0$, we have

$$g'(t) = \int_0^t K(r+s,s)f'(t-s)ds + K(r+t,t)f(0).$$

On the other hand,

$$\begin{aligned}
\frac{g(t+h)-g(t)}{h} &= \frac{1}{h}\Big[\int_0^{t+h} K(r+s,t+h-s)f(s)ds - \int_0^t K(r+s,t-s)f(s)ds\Big] \\
&= \frac{1}{h}\Big[\int_0^{t} K(r+s,t+h-s)f(s)ds - \int_0^t K(r+s,t-s)f(s)ds \\
&+ \int_t^{t+h} K(r+s,t+h-s)f(s)ds\Big].
\end{aligned}$$

By the identities (2.2)

$$\begin{aligned}
K(r+s,t+h-s)f(s) &= K(r+s+t+h-s-h,h)K(r+s,t+h-s-h)f(s) \\
&= K(r+t,h)K(r+s,t-s)f(s).
\end{aligned}$$

Hence

$$\begin{aligned}
\frac{g(t+h)-g(t)}{h} &= \frac{1}{h}\Big[\int_0^{t} K(r+t,h)K(r+s,t-s)f(s)ds \\
&- \int_0^t K(r+s,t-s)f(s)ds + \int_t^{t+h} K(r+s,t+h-s)f(s)ds\Big] \\
&= \frac{1}{h}\Big[\int_0^{t} K(r+s,t-s)\Big(K(r+t,h)f(s)-f(s)\Big)ds \\
&+ \int_t^{t+h} K(r+s,t+h-s)f(s)ds\Big].
\end{aligned}$$

Passing to the limit when $h \to 0$,

$$\begin{aligned}
g'(t) &= \int_0^t K(r+s,t-s)A(r+t)f(s)ds + K(r+t,t-t)f(t) \\
&= \int_0^t K(r+s,t-s)A(r+t)f(s)ds + f(t),
\end{aligned}$$

thus $g'(t)$ exists. Also,

$$\int_0^t K(r+s,t-s)A(r+t)f(s)ds = A(r+t)\int_0^t K(r+s,t-s)f(s)ds.$$

and

$$g'(t) = A(r+t)g(t) + f(t), \qquad (t \geq 0).$$

Proving that g is strongly differentiable and the equation (2.7) is solution of the initial value problem (2.6). Indeed,

$$\begin{aligned}
\dot{x}(t) &= \frac{\partial K(r,t)x_0}{\partial t} + A(r+t)\int_0^t K(r+s,t-s)f(s)ds + f(t) \\
&= A(r+t)K(r,t)x_0 + A(r+t)\int_0^t K(r+s,t-s)f(s)ds + f(t) \\
&= A(r+t)\Big(K(r,t)x_0 + \int_0^t K(r+s,t-s)f(s)ds\Big) + f(t) \\
&= A(r+t)x(t) + f(t).
\end{aligned}$$

The uniqueness is consequence of Theorem 2.2

Definition 2.3 Let $f \in L_p(0,T;X)$, $1 \leq p < \infty$, $x_0 \in X$, and $K(r,t)$ is a strongly continuous quasisemigroup. The function

$$x_r(t) = K(r,t)x_0 + \int_0^t K(r+s,t-s)f(s)ds \tag{2.10}$$

is defined to be the mild solution of the equation (2.6) on $[0,T]$.

The function $x_r(t)$ is clearly continuous.

Proposition 2.3 *Let $A(t)$ with domain $D(A(t))$ be the generator of a strongly continuous quasisemigroup $K(t,s)$ on a Banach space X and $B \in L(X)$, such that $B^2 = B$ with $BK(t,s) = K(t,s)B$, $\forall s,t \geq 0$. Then $A(t)B$ is the generator of the strongly continuous quasisemigroup*

$$R(t,s) = B[K(t,s) - I] + I.$$

Proof. 2.6 The proof follows directly from the definition of $R(t,s)$ and the condition imposed to B.

Proposition 2.4 *Let $A(t)$ and $K(t,s)$ be as in the Proposition 2.3 and $B \in L(X)$ such that $BK(t,s) = K(t,s)B$, $\forall s,t \geq 0$. Then $A(t) + B$ is the generator of the strongly continuous quasisemigroup defined by*

$$R(t,s) = e^{sB}K(t,s).$$

Proof. 2.7 The proof follows directly from the definition of $R(t,s)$ and the condition imposed to B.

Proposition 2.5 *Let $A(t)$ with domain $D(A(t))$ be the generator of a strongly continuous quasisemigroup $K(r,t)$ on a Banach space X and $B \in L(X)$ such that $BK(r,t) = K(r,t)B$, $\forall r,t \geq 0$. Then, the function $x_r(t) = R(r,t)x_0$ is the solution of the integral equation*

$$z(t) = K(r,t)x_0 + \int_0^t K(r+s,t-s)Bz(s)ds. \tag{2.11}$$

Proof. 2.8 ¿From Proposition 2.4, we have

$$\begin{aligned} x_r(t) &= R(r,t)x_0 = K(r,t)e^{tB}x_0 = K(r,t)x_0 + K(r,t)e^{tB}x_0 - K(r,t)x_0 \\ &= K(r,t)x_0 + K(r,t)\int_0^t Be^{sB}x_0 ds \\ &= K(r,t)x_0 + \int_0^t K(r+s,t-s)K(r,s)e^{sB}Bx_0 ds \\ &= K(r,t)x_0 + \int_0^t K(r+s,t-s)BK(r,s)e^{sB}x_0 ds \\ &= K(r,t)x_0 + \int_0^t K(r+s,t-s)Bx_r(s)ds. \end{aligned}$$

Thus, the function $x_r(t)$ is solution of the equation (2.11).

Example 2.5 Let $r \geq 0$, $\mu \in \mathbb{R}$ and consider the problem

$$P_{\mu,r}\Big) \quad \begin{cases} \frac{\partial x(t,\xi)}{\partial t} = 2(r+t)\frac{\partial x(t,\xi)}{\partial \xi} + \mu x(t,\xi) \\ x(0,\xi) = x_0(\xi), \qquad (\xi, t \geq 0). \end{cases} \tag{2.12}$$

Let X be the Banach space of bounded uniformly continuous defined in $[0,\infty)$ with the of supremum norm, D the dense subspace on X given by $D = \{x \in X : \dot{x} \in X\}$ and $K(t,s)x(\xi) = x(s^2 + 2st + \xi)$. Then, the generator $A(t)$ of $K(t,s)$ is give by

$$\begin{aligned} A(t): \ D &\longrightarrow X \\ \phi &\longrightarrow A(t)\phi = 2t\dot{\phi} \quad (t \geq 0). \end{aligned}$$

If we define $B : D \to X$ by $Bx = \mu x$, then $BK(t,s) = K(t,s)B$.
So, equation (2.12) can be written as

$$P_r\Big) \quad \begin{cases} \dot{x}(t) = \Big(A(r+t) + B\Big)x(t), \quad (t > 0) \\ x(0) = x_0. \end{cases} \tag{2.13}$$

If $x_0 \in D$, then by Theorem 2.2 and Proposition 2.4 we obtain that

$$x(t) = R(r,t)x_0 = e^{Bt}K(r,t)x_0$$

is the only solution of the equation (2.13).
Consequently, the problem $P_{\mu,r}$ admits as unique solution the function

$$x(t,\xi) = e^{\mu t}x_0(t^2 + 2rt + \xi).$$

Example 2.6 Let us consider the problem

$$P_\mu\Big) \quad \begin{cases} \frac{\partial x(t,\xi)}{\partial t} = \frac{\partial x(t,t+\xi)}{\partial \xi} + \mu x(t,\xi) \\ x(0,\xi) = x_0(\xi), \qquad (t, \xi \geq 0), \end{cases} \tag{2.14}$$

where $\mu \in \mathbb{R}$ is fixed.

Using the Example 2.3 and the Proposition 2.4, we obtain that the only solution of P_μ), is

$$x(t,\xi) = e^{\mu t - 1}\sum_{n=0}^{\infty} \frac{x_0(nt+\xi)}{n!}, \quad (x_0 \in D).$$

In fact, if X and D are as in Example 2.5, then

$$\begin{aligned} A : D \subseteq X &\longrightarrow X \\ \phi &\longrightarrow A\phi = \dot{\phi} \end{aligned}$$

generates a strongly continuous semigroup given by

$$T_t\phi(\xi) = \phi(t+\xi).$$

So the problem (2.14) can be expressed as

$$P_{\mu'}\Big) \qquad \begin{cases} \dot{x}(t) = AT_t x(t) + B_\mu x(t) = \Big(A(t) + B_\mu\Big)x(t) \\ \\ x(0) = x_0 \in D, \qquad t \geq 0, \end{cases} \tag{2.15}$$

where

$$B_\mu = \mu I \qquad \textit{and} \qquad A(t) = AT_t : D \longrightarrow X.$$

$A(t)$ is the generator of the quasisemigroup given by

$$K(t,s) = e^{T_{t+s}-T_t}.$$

Therefore, the operator $A(t) + B_\mu$ generates the quasisemigroup $R(t,s)$ given by

$$R(t,s) = e^{sB_\mu}K(t,s).$$

Then, the solution of the initial value problem $P_{\mu'})$ can be obtained from

$$x(t) = R(0,t)x_0.$$

In fact,

$$\begin{aligned} x(t) &= e^{\mu t}K(0,t)x_0 = e^{\mu t}e^{T_t - I}x_0 \\ &= e^{\mu t}e^{-1}e^{T_t}x_0 = e^{\mu t-1}\sum_{n=0}^{\infty}\frac{T_t^n x_0}{n!} \\ &= e^{\mu t-1}\sum_{n=0}^{\infty}\frac{T_{nt}x_0}{n!}. \end{aligned}$$

Evaluating $x(t)(\cdot)$ in ξ, we obtain that

$$x(t,\xi) = e^{\mu t-1}\sum_{n=0}^{\infty}\frac{x_0(nt+\xi)}{n!}.$$

The following Theorem is a generalization of the Proposition 2.4.

Theorem 2.4 *Let $A(t)$ be the generator of the strongly continuous quasi-semigroup $K(t,s)$ on a Banach space X and $B \in L(X)$. Then $A(t) + B$ is the generator of a strongly continuous quasi-semigroup $R(t,s)$ defined by*

$$R(r,t)x_0 = K(r,t)x_0 + \int_0^t K(r+s, t-s)BR(r,s)x_0 ds. \tag{2.16}$$

Moreover, if

$$\|K(r,t)\| \leq M(r+t),$$

then

$$\|R(r,t)\| \leq M(r+t)e^{\|B\|M(r+t)t}.$$

Proof. 2.9 The solution of the equation (2.16) is given by

$$R(r,t) = \sum_{n=0}^{\infty} R_n(r,t),$$

where

$$R_0(r,t) = K(r,t)$$

and

$$R_n(r,t)x_0 = \int_0^t K(r+s,t-s)BR_{n-1}(r,s)x_0 ds, \quad n = 1,2,3\cdots. \tag{2.17}$$

First, we prove the uniform convergence on compact set.
Since $K(r,t)$ is a quasisemigroup, for $n = 0$ we have that

$$\|R_0(r,t)\| \leq M(r+t).$$

For $n = 1$,

$$R_1(r,t)x_0 = \int_0^t K(r+s,t-s)BR_0(r,s)x_0 ds.$$

Hence

$$\begin{aligned} \|R_1(r,t)x_0\| &\leq \int_0^t \|K(r+s,t-s)\|\|B\|\|R_0(r,s)\|\|x_0\|ds \\ &\leq M(r+t)\|B\|M(r+t)\|x_0\| \int_0^t ds \\ &= M(r+t)\|B\|M(r+t)t\|x_0\|. \end{aligned}$$

Proceeding inductively we see that

$$\|R_n(r,t)x_0\| \leq M(r+t)\frac{\Big[\|B\|M(r+t)t\Big]^n}{n!}\|x_0\|, \quad \forall n \geq 0 \quad and \quad \forall x_0 \in X.$$

Thus,

$$\|R_n(r,t)\| \leq M(r+t)\frac{\Big[\|B\|M(r+t)t\Big]^n}{n!}, \quad \forall n \geq 0.$$

Then,

$$R(r,t) = \sum_{n=0}^{\infty} R_n(r,t)$$

is bounded by the convergent series

$$M(r+t)\sum_{n=0}^{\infty}\frac{\Big[\|B\|M(r+t)t\Big]^n}{n!} = M(r+t)e^{\|B\|M(r+t)t}.$$

Therefore, the series

$$R(r,t)=\sum_{n=0}^{\infty}R_n(r,t)$$

converges uniformly on compact sets.
Now, we will see that $R(r,t)x_0$ is solution of (2.16).

$$\begin{aligned}
R(r,t)x_0 &= \sum_{n=0}^{\infty}R_n(r,t)x_0=R_0(r,t)x_0+\sum_{n=1}^{\infty}R_n(r,t)x_0\\
&= K(r,t)x_0+\sum_{n=1}^{\infty}\int_0^t K(r+s,t-s)BR_{n-1}(r,s)x_0ds\\
&= K(r,t)x_0+\int_0^t K(r+s,t-s)B\sum_{n=0}^{\infty}R_n(r,s)x_0ds\\
&= K(r,t)x_0+\int_0^t K(r+s,t-s)BR(r,s)x_0ds.
\end{aligned}$$

Thus, for each fixed $r\geq 0$, $R(r,t)x_0$ is solution of the equation (2.16).
$R(r,t)x_0$ is the only solution of (2.16). In fact, if $x_r(t)$ is another solution of the equation (2.16), then

$$\begin{aligned}
\|R(r,t)x_0-x_r(t)\| &= \|K(r,t)x_0+\int_0^t K(r+s,t-s)BR(r,s)x_0ds\\
&- K(r,t)x_0-\int_0^t K(r+s,t-s)Bx_r(s)ds\|\\
&\leq \int_0^t \|K(r+s,t-s)\|\|B\|\|R(r,s)x_0-x_r(s)\|ds\\
&\leq \int_0^t M(r+t)\|B\|\|R(r,s)x_0-x_r(s)\|ds.
\end{aligned}$$

By Gronwall's lemma, we have that

$$\|R(r,t)x_0\quad x_r(t)\|\leq 0e^{\int_0^t M(r+t)\|B\|ds},$$

therefore

$$R(r,t)x_0=x_r(t),\qquad (t\geq 0).$$

Now, we will prove that $R(r,t)$ satisfies the condition of the Definition 2.1.
Clearly 1) is satisfied.

Now, we shall consider the following estimate,

$$R(r,t)x_0-R(r,t_0)x_0.$$

We have

$$\begin{aligned}
\|R(r,t)x_0-R(r,t_0)x_0\| &\leq \|K(r,t)x_0-K(r,t_0)x_0\|\\
&+ \int_{t_0}^t |K(r+s,t-s)\|\|B\|\|R(r,s)x_0\|ds\\
&+ \|\int_0^{t_0}[K(r+s,t-s)-K(r+s,t_0-s)]BR(r,s)x_0ds\|.
\end{aligned}$$

which goes to a zero as $t \longrightarrow t_0$. Therefore $R(r,\cdot)$ is strongly continuous. Now we will prove that $R(\cdot,s)$, is strongly continuous.

$$\begin{aligned}
\|R(r,s)x_0 - R(r_0,s)x_0\| &= \|K(r,s)x_0 \\
&+ \int_0^s K(r+\alpha,s-\alpha)BR(r,\alpha)x_0 d\alpha - K(r_0,s)x_0 \\
&- \int_0^s K(r_0+\alpha,s-\alpha)BR(r_0,\alpha)x_0 d\alpha\| \\
&\leq \|K(r,s)x_0 - K(r_0,s)x_0\| \\
&+ \int_0^s \|K(r+\alpha,s-\alpha)BR(r_0,\alpha)x_0 \\
&- K(r_0+\alpha,s-\alpha)BR(r_0,\alpha)x_0\|d\alpha \\
&+ \int_0^s \|K(r+\alpha,s-\alpha)\|\|B\|\|R(r,\alpha)x_0 - R(r_0,\alpha)x_0\|d\alpha.
\end{aligned}$$

Then,

$$\begin{aligned}
\|R(r,s)x_0 - R(r_0,s)x_0\| &\leq \|K(r,s)x_0 - K(r_0,s)x_0\| \\
&+ \int_0^s \|K(r+\alpha,s-\alpha) \\
&- K(r_0+\alpha,s-\alpha)\|\|B\|\|R(R_0,\alpha)X_0\|d\alpha \\
&+ \int_0^s M(r+s)\|B\|\|R(r,\alpha)x_0 - R(r_0,\alpha)x_0\|d\alpha.
\end{aligned}$$

Since $K(\cdot,\cdot)$ is a quasisemigroup, using parts 3) and 4) of Definition 2.1 and the dominated convergence theorem, we obtain that the first two terms go to zero.

Therefore

$$\|R(r,s)x_0 - R(r_0,s)x_0\| \leq \epsilon + \int_0^s M(r+s)\|B\|\|R(r,\alpha)x_0 - R(r_0,\alpha)x_0\|d\alpha.$$

As a consequence of the strong continuity of $R(r,\cdot)$ and by Gronwall's lemma, it follows that

$$\|R(r,s)x_0 - R(r_0,s)x_0\| \leq \epsilon\, e^{\|B\|M(r+s)s},$$

which implies that $R(\cdot, s)$ is strongly continuous. Now,

$$\begin{aligned}
\|R(r,t)x_0 - R(r_0,t_0)x_0\| &= \|K(r,t)x_0 + \int_0^t K(r+s,t-s)BR(r,s)x_0 ds \\
&- K(r_0,t_0)x_0 - \int_0^{t_0} K(r_0+s,t_0-s)BR(r_0,s)x_0 ds\| \\
&\leq \|K(r,t)x_0 - K(r_0,t_0)x_0\| \\
&+ \|\int_0^{t_0} K(r+s,t-s)B[R(r,s)x_0 - R(r_0,s)x_0]ds\| \\
&+ \|\int_0^{t_0} [K(r+s,t-s)BR(r_0,s)x_0 \\
&- K(r_0+s,t_0-s)BR(r_0,s)x_0]ds\| \\
&+ \|\int_{t_0}^t K(r+s,t-s)BR(r,s)x_0 ds\|.
\end{aligned}$$

Taking limits when (r,t) goes to (r_0,t_0), the right side of the last inequality converges to zero, which proves 3) of Definition 2.1.
To prove 2) of Definition 2.1 we consider the equalities

$$z(s) = R(r,t+s)x_0 \qquad \textit{and} \qquad w(s) = R(r+t,s)R(r,t)x_0,$$

for $t > 0$ fixed.
Plainly, z and w satisfies

$$z(0) = R(r,t)x_0 \qquad \textit{and} \qquad w(0) = R(r+t,0)R(r,t)x_0 = R(r,t)x_0.$$

Furthermore

$$z(s) = R(r,t+s)x_0 = K(r,t+s)x_0 + \int_0^{t+s} K(r+\alpha,t+s-\alpha)BR(r,\alpha)x_0 d\alpha.$$

Having the change $u = \alpha - t$, we have

$$\begin{aligned}
z(s) &= k(r,t+s)x_0 + \int_{-t}^s K(r+t+u,s-u)BR(r,t+u)x_0 du \\
&= k(r,t+s)x_0 + \int_{-t}^0 K(r+t+u,s-u)Bz(u)du \\
&+ \int_0^s K(r+t+u,s-u)Bz(u)du.
\end{aligned}$$

Observes that

$$z(0) = K(r,t)x_0 + \int_{-t}^0 K(r+t+u,-u)Bz(u)du.$$

We still have to prove that

$$K(r,t)x_0 + \int_{-t}^0 K(r+t+u,s-u)Bz(u)du = R(r,t)x_0.$$

by the same change of variable we have

$$\begin{aligned}\int_{-t}^{0} K(r+t+u,s-u)Bz(u)du &= \int_{0}^{t} K(r+\alpha,t+s-\alpha)Bz(\alpha-t)d\alpha \\ &= \int_{0}^{t} K(r+\alpha,t+s-\alpha)BR(r,\alpha)x_0 d\alpha.\end{aligned}$$

So

$$\begin{aligned}z(s) &= K(r,t+s)x_0 + \int_0^t K(r+\alpha,t+s-\alpha)BR(r,\alpha)x_0 d\alpha \\ &+ \int_0^s K(r+t+\alpha,s-\alpha)Bz(\alpha)d\alpha \\ &= K(r+t,s)K(r,t)x_0 + \int_0^t K(r+t,s)K(r+\alpha,t-\alpha)BR(r,\alpha)d\alpha \\ &+ \int_0^s K(r+t+\alpha,s-\alpha)Bz(\alpha)d\alpha \\ &= K(r+t,s)R(r,t)x_0 + \int_0^s K(r+t+\alpha,s-\alpha)Bz(\alpha)d\alpha\end{aligned}$$

is solution of the equation (2.16).
On the other hand,

$$\begin{aligned}w(s) &= R(r+t,s)R(r,t)x_0 = K(r+t,s)R(r,t)x_0 \\ &+ \int_0^s K(r+t+\alpha,s-\alpha)BR(r+t,\alpha)R(r,t)x_0 d\alpha \\ &= K(r+t,s)R(r,t)x_0 + \int_0^s K(r+t+\alpha,s-\alpha)Bw(\alpha)d\alpha.\end{aligned}$$

which implies that w is another solution of the equation (2.16). From the uniqueness of the solution we have

$$z(s) = w(s) \Longrightarrow R(r,t+s)x_0 = R(r+t,s)R(r,t)x_0,$$

which proves part 2 of the definition 2.1.
Lets $t \geq 0$, $s > 0$ and $x_0 \in D$,

$$\begin{aligned}\lim_{s\to 0^+} \frac{R(t,s)x_0 - x_0}{x_0} &= \lim_{s\to 0^+} \frac{1}{s}\Big[K(t,s)x_0 + \int_0^s K(t+\alpha,s-\alpha)BR(t,\alpha)x_0 d\alpha - x_0\Big] \\ &= \lim_{s\to 0^+} \frac{K(t,s)x_0 - x_0}{s} \\ &+ \lim_{s\to 0^+} \int_0^s K(t+\alpha,s-\alpha)BR(t,\alpha)x_0 d\alpha \\ &= A(t)x_0 + K(t,0)BR(t,0)x_0 = (A(t)+B)x_0.\end{aligned}$$

For $t > s > 0$,

$$\begin{aligned}\lim_{s\to 0^+}\frac{R(t-s,s)x_0-x_0}{s} &= \lim_{s\to 0^+}\Big[\frac{K(t-s,s)x_0-x_0}{s}\\ &+ \frac{1}{s}\int_0^s K(t-s+\alpha,s-\alpha)BR(t-s,\alpha)x_0d\alpha\Big]\\ &= A(t)x_0+K(t,0)BR(t,0)x_0=(A(t)+B)x_0.\end{aligned}$$

Since D is a dense on X, then

$$A(t)+B$$

is the generator of $R(t,s)$.

References

[1] Barcenas D. and Leiva H.. "Sobre La Existencia Y La Unicidad De Las Soluciones De Una Clase De Ecuaciones Diferenciales En Espacios De Banach". *Notas de Matemáticas N 87*, Universidad de Los Andes, Mérida, Venezuela (1987).

[2] Barcenas D. and Leiva H.. Quasisemigroups, Evolutions Equation and Controllability. *Notas de Matemáticas N 109*, Universidad de Los Andes, Mérida, Venezuela (1991).

[3] Cuc V. On the exponentially bounded C_0-quasisemigroups, *Analele Universitatii din Timisoara, Seria Matematica - Informatica* **35** (2) (1997),193-199.

[4] Cuc V. Non-uniform exponential stability of C_0-quasisemigroups in Banach space, *Radovi Matematichi*, Vol. 10 (2001), 35-45.

[5] Cuc V. A Generalization of a theorem of Datko and Pazy New Zealand *Journal of Mathematics*. Volume 30 (2001), 31-39.

[6] Curtain R.F. and Pritchard A.J. Infinite Dimensional Linear Systems, *Lecture Notes in Control and Information Sciences*, Vol. 8. Springer Verlang, Berlin. Heidelberg. New York, (1978).

[7] Dunford N. and Schwartz J.. *Linear Operators*, Vol. I, Interscience, (1958).

[8] Engel Klaus Jochen and Nagel Rainer. *One-Parameter Semigroups for Linear Evolution Equations*, Springer-Verlag, Berlin. Heidelberg. New York, (2000).

[9] Hille E. and Phillips R.S. Functional Analysis and semigroups, *Amer. Math. Soc. Coll. Publ.*, **31**, (1957)

[10] Kato T. Integration of equation of evolution in Banach Space, *J. Math. Soc. Japan*, **5**,208-234, (1953).

[11] Ladas C.E. and Lakshmikantham. *Differential Equations in Abstract Spaces*, Vol. 85, Academic Press, New York,(1972).

[12] Neerven Jan Van. *The Adjoint of a Semigroup of Linear Operators*. Springer-Verlag, Berlin. Heidelberg. New York, (1992).

[13] Pazy A. Semigroups of Linear Operators and Applications to Partial Differential Equations, A*pplied Mathematical Sciences*, Vol. 44, Springer-Verlag, Berlin. Heidelberg. New York, (1983).

[14] Yosida Kôsaku. *Functional Analysis*, fourth edition, Springer-Verlag, Berlin. Heidelberg. New York, (1974).

In: Focus on Evolution Equations
Editor: Gaston M. N'Guerekata

ISBN: 978-1-60021-342-7

Chapter 13

CONTINUATION AND MAXIMAL REGULARITY OF FRACTIONAL-ORDER EVOLUTIONARY INTEGRAL EQUATION

Ahmed M.A. El-Sayed[1,*] ***and Mohamed A.E. Herzallah***[2,†]
[1]Faculty of Science, Alexandria University, Alexandria, Egypt
[2]Faculty of Science, Zagazig University, Zagazig, Egypt

Abstract

The maximal regularity and continuation of the solution and its fractional-order derivative of the solution of the Cauchy problem of the non-homogeneous fractional order evolution equation $D^{\alpha}u(t) = Au(t) + f(t),\ \alpha \in (0,1)$ have been studied in [6]. Here we study the maximal regularity, continuation of the solution and its derivative and some other properties of the solution of the Cauchy problem of the non-homogeneous fractional order evolutionary integral equation
$D^{\alpha}u(t) = \int_0^t h(t-s)Au(s)ds + f(t),\ \alpha \in (0,1)$.

Keywords and phrases: Evolutionary integral equation, fractional order derivative, maximal regularity.

1. Introduction

Let X be a Banach space with norm $\|.\|$. We shall define some classes of functions u from the interval $J = [0,T]$, $T < \infty$, of R into X which will be used in this paper. Let $L(J,X)$ be the space of integrable functions defined on the interval J with values in X, $C(J,X)$ be the space of continuous functions from J to X with the sup-norm, $C^m(J,X) = \{u \in C(J,X):\ D^k u \in C(J,X),\ k \le m\}$ and $W^{m,p}(J,X)$ is the Sobolev spaces.
Finally for each $\theta \in (0,1)$ we define the space of Hölder continuous functions

$$C^{\theta}(J,X) = \{u : J \to X:\ \|u\|_{\theta} = \sup_{t,s \in J} \frac{\|u(t)-u(s)\|}{|t-s|^{\theta}} < \infty\}$$

*E-mail address: amasayed@hotmail.com
†E-mail address: m_herzallah75@hotmail.com

with norm $\|u\|_{C^\theta(J,X)} = \|u\|_{C(J,X)} + \|u\|_\theta$. Finally define

$$C_0^\theta(J,X) = \{u \in C^\theta(J,X) : \ u(0) = 0\}.$$

Let A be a closed linear operator with dense domain $D(A) \subset X$. It is known (see [14]) $X_A = D(A)$ is a Banach space under the graph norm $\|x\|_A = \|x\| + \|Ax\|$.
Consider the non-homogeneous evolution equation Cauchy problem;

$$D^\alpha u(t) = Au(t) + f(t), \quad t \in (0,T], \text{ with } u(0) = u_o, \ \alpha \in (0,1] \tag{1}$$

where D^α is the Caputo derivative (see Definition 2.2). The authors (see [6]) proved, under certain conditions, that problem (1) has a unique strong solution $u_\alpha \in C(J; X_A)$ if $f \in W^{1,1}(J; X_A)$ and a unique mild solution $u_\alpha \in C(J;X)$ if $f \in W^{1,1}(J;X)$. Also they proved the continuation of the strong solution and its fractional derivative as $\alpha \to 1^-$ (i.e. $\lim_{\alpha\to 1^-} u_\alpha(t) = u_1(t)$ and $\lim_{\alpha \to 1^-} D^\alpha u_\alpha(t) = Du_1(t)$ where $u_1(t)$ is the solution of the evolution equation $Du(t) = Au(t) + f(t)$) and studied the maximal regularity property for problem (1) with $u(0) = 0$ by proving that if $f \in C_0^\delta(J,X)$ for $\delta \in (0,1)$ then there is a strong solution $u_\alpha \in C_0^\delta(J, X_A)$ with $D^\alpha u_\alpha \in C_0^\delta(J,X)$. Also the two authors (see [5]) studied the continuation of the solution and the fractional derivative of it of the homogeneous fractional order evolutionary integral equation;

$$D^\gamma u(t) = \int_0^t h(t-s)Au(s)ds, \quad u(0) = u_o.$$

and studied some properties of the solution and some applications as special cases of this problem.
Here we study the existence and uniqueness of the solution of the Cauchy problem of the non-homogeneous fractional order evolutionary integral equation

$$D^\alpha u(t) = f(t) + \int_0^t h(t-s)Au(s)ds, \quad u(0) = u_o, \alpha \in (0,1) \qquad (2).$$

The continuation properties of the solution $u_\alpha(t)$ and its fractional order derivative $D^\alpha u_\alpha(t)$ will be proved. Finally we study the maximal regularity property.

2. Preliminaries

Let $f \in L^1(J,R)$ and let α be a positive real number.
Definition 2.1 The fractional integral of order α of the function $f(t)$ is defined by (see [8],[11],[12],[13] and [15])

$$I^\alpha f(t) = \int_0^t \frac{(t-s)^{\alpha-1}}{\Gamma(\alpha)} f(s)ds = f(t) * \phi_\alpha(t) \tag{3}$$

where $\phi_\alpha(t) = \frac{t^{\alpha-1}}{\Gamma(\alpha)}$ for $t > 0$ and $\phi_\alpha(t) = 0$ for $t \leq 0$, and (see [7]) $\phi_\alpha(t) \to \delta(t)$ (the delta function) as $\alpha \to 0$.

Definition (Caputo derivative) 2.2 The fractional derivative of order $\alpha \in (0,1)$ of the function $g(t)$ is defined by (see [1-6, 8, 12, 13])

$$D^{\alpha}g(t) = I^{1-\alpha}Dg(t), \qquad D = \frac{d}{dt}. \tag{4}$$

Consider now the integral equation

$$u(t) = f(t) + \int_0^t a(t-s)Au(s)ds \tag{5}$$

where A is a closed linear unbounded operator with dense domain $D(A) = X_A$ in the Banach space X and $0 \neq a(t) \in L^1_{loc}(R^+)$. The existence of a unique solution of equation (5) and its properties have been studied by some authors (see [9,14]). Till the end of this section we state some definitions and theorems (see [14]) which are used in the recent.
Definition 2.3 A function $u \in C(J;X)$ is called

(a) a strong solution of (5) on J if $u \in C(J;X_A)$ and (5) holds on J;

(b) a mild solution of (5) on J if $a * u \in C(J;X_A)$ and $u(t) = f(t) + A(a*u)(t)$ on J.

Definition 2.4 A family $S(t)_{t\geq 0} \subset B(X)$ of bounded linear operators in X is called a resolvent for (5) if the following conditions are satisfied,

(1) $S(t)$ is strongly continuous on R^+ and $S(0) = I$;

(2) $S(t)$ commutes with A, which means that $S(t)D(A) \subset D(A)$ and $AS(t)x = S(t)Ax$ for all $x \in D(A)$ and $t \geq 0$;

(3) the resolvent equation

$$S(t)x = x + \int_0^t a(t-s)AS(s)xds \quad \text{for all} \quad x \in D(A), \quad t \geq 0 \tag{6}$$

holds.
Proposition 2.1 The evolutionary integral equation (5) admits at most one resolvent $S(t)$.
Proposition 2.2 Suppose (5) admits a resolvent $S(t)$ and let $f \in C(J;X)$.

(i) If $f \in W^{1,1}(J;X)$, then

$$u(t) = S(t)f(0) + \int_0^t S(t-s)\dot{f}(s)ds, \quad t \in J \tag{7}$$

is a mild solution of (5);

(ii) If $f \in W^{1,1}(J;X_A)$, then $u(t)$ defined by (7) is a strong solution of (5).

Definition 2.5 Equation (5) with $a \in L^1_{loc}(R^+)$ be of sub exponential growth, which means $\int_0^\infty e^{-\epsilon t}|a(t)|dt < \infty$ for all $\epsilon > 0$, is called "Parabolic" if the following conditions hold:

(1) $\hat{a}(\lambda) \neq 0$, $1/\hat{a}(\lambda) \in \rho(A)$ for all $Re(\lambda) > 0$;

(2) There is a constant $M \geq 1$ such that $H(\lambda) = (I - \hat{a}(\lambda)A)^{-1}/\lambda$ satisfies

$$\|H(\lambda)\| \leq \frac{M}{|\lambda|} \quad \text{for all} \quad Re(\lambda) > 0$$

where $\hat{a}(\lambda)$ is the Laplace transform of $a(t)$.

Definition 2.6 Let $a \in L^1_{loc}(R^+)$ be of sub exponential growth and suppose $\hat{a}(\lambda) \neq 0$ for all $Re(\lambda) > 0$. a is called sectorial with angle $\theta > 0$(or merely θ-sectorial) if $|arg(\hat{a}(\lambda)| \leq \theta$ for all $Re(\lambda) > 0$.

Definition 2.7 Let $a \in L^1_{loc}(R^+)$ be of sub exponential growth and $k \in N$. $a(t)$ is called k-regular if there is a constant $c > 0$ such that

$$|\lambda^n \hat{a}^{(n)}(\lambda)| \leq c|\hat{a}(\lambda)| \quad \text{for all} \quad Re(\lambda) > 0, \quad 0 \leq n \leq k.$$

Theorem 2.1 Let X be a Banach space, A is a closed linear operator in X with dense domain $D(A)$ and $a \in L^1_{loc}(R^+)$. Assume (5) is parabolic and $a(t)$ is k-regular, for some $k \geq 1$. Then there is a resolvent $S \in C^{k-1}((0, \infty); B(X))$ for (5) and there is a constant $M \geq 1$ such that the estimates

$$\|t^n S^{(n)}(t)\| \leq M \quad \text{for all} \quad t \geq 0, \quad n \leq k-1 \tag{8}$$

$$\|t^k S^{(k-1)}(t) - s^k S^{(k-1)}(s)\| \leq M(t-s)(1 + log(\frac{t}{t-s})), \quad 0 \leq s < t < \infty \tag{9}$$

are valid.

Corollary 2.1 Let $a \in L^1_{loc}(R^+)$ be 1-regular and sectorial with $\theta < \frac{\pi}{2}$ and suppose A generates a bounded C_o-semigroup $\{T(t), \ t \geq 0\}$ in the Banach space X. Then (5) is parabolic and there exists a bounded resolvent $S(t)$ for it.

Theorem 2.2 Suppose $a \in L^1_{loc}(R^+)$ is 2-regular and such that (5) is parabolic and let $\alpha \in (0, 1)$. Then

(i) for each $f \in C^\alpha_0(J; X)$ there is a mild solution $u \in C^\alpha_0(J; X)$ to (5);

(ii) for each $f \in C^\alpha_0(J; X_A)$ there is a strong solution $u \in C^\alpha_0(J; X_A)$ to (5);

3. Existence of Solution

Consider the Cauchy problem (2) where A is a closed linear operator with dense domain $D(A) = X_A \subset X$.

Definition 3.1 A function $u \in C(J; X)$ is called

(a) a strong solution of (2) on J if $u \in C(J; X_A)$ and (2) holds on J;

(b) a mild solution of (2) on J if $h_\alpha * u \in C(J; X_A)$, $h_\alpha(t) = \phi_\alpha(t) * h(t)$, and $u(t) = u_o + (\phi_\alpha * f)(t) + A((h_\alpha * u)(t))$ on J.

Theorem 3.1 Let $\alpha \in (0, 1), u_o \in D(A), f \in W^{1,1}(J, X_A)$ with $h(t)$ is 2-regular and A is the infinitesimal generator of a bounded C_o semigroup $\{T(t), \ t \geq 0\}$.

If $|arg(\frac{\hat{h}(\lambda+1)}{(\lambda+1)^{\alpha}}| \leq \theta < \frac{\pi}{2}$ then there is a unique strong solution $u_{\alpha} \in C(J, X_A) \cap C^1((0,T]; X)$ of (2) given by

$$u_{\alpha}(t) = e^t S_{\alpha}(t) u_o - e^t S_{\alpha}(t) * u_o + e^t S_{\alpha}(t) * \phi_{\alpha}(t) * (f(0)\delta(t) + \dot{f}(t) - f(t))$$

where $S_{\alpha} \in C^1(J; B(X))$ is a bounded resolvent given as in Definition 2.4, with $a(t) = H_{\alpha}(t)$, and satisfies Theorem 2.1.

Proof. Operating by the convolution of $\phi_{\alpha}(t)$ on (2) we get the integral equation

$$u_{\alpha}(t) = f_{\alpha}(t) + \int_0^t h_{\alpha}(t-s) A u_{\alpha}(s) ds \qquad (10)$$

where $f_{\alpha}(t) = u_o + \phi_{\alpha}(t) * f(t)$ and $h_{\alpha}(t) = \phi_{\alpha}(t) * h(t)$. If we let $e^{-t} u_{\alpha}(t) = v_{\alpha}(t)$ we get

$$v_{\alpha}(t) = F_{\alpha}(t) + H_{\alpha}(t) * A v_{\alpha}(t) \qquad (11)$$

where $F_{\alpha}(t) = e^{-t} f_{\alpha}(t)$ and $H_{\alpha}(t) = e^{-t} h_{\alpha}(t)$.

Now we apply the conditions of Theorem 2.1 and Corollary 2.1

$$H_{\alpha}(t) = e^{-t} h_{\alpha}(t) \Rightarrow \hat{H}_{\alpha}(\lambda) = \frac{1}{(\lambda+1)^{\alpha}} \hat{h}(\lambda+1)$$

$$\Rightarrow \hat{H}'_{\alpha}(\lambda) = \frac{-\alpha}{(\lambda+1)^{\alpha+1}} \hat{h}(\lambda+1) + \frac{1}{(\lambda+1)^{\alpha}} \hat{h}'(\lambda+1)$$

and where $h(t)$ is 2-regular we get

$$\begin{aligned}
|\lambda \hat{H}'_{\alpha}(\lambda)| &\leq |\frac{\lambda\alpha}{(\lambda+1)^{\alpha+1}}||\hat{h}(\lambda+1)| + |\frac{\lambda}{(\lambda+1)^{\alpha}}||\hat{h}'(\lambda+1)| \\
&\leq |\frac{\hat{h}(\lambda+1)}{(\lambda+1)^{\alpha}}|\{|\frac{\alpha\lambda}{\lambda+1}| + c|\frac{\lambda}{\lambda+1}|\} \\
&\leq (\alpha+c)|\hat{H}_{\alpha}(\lambda)| \\
&\leq c_1 |\hat{H}_{\alpha}(\lambda)|, \qquad c_1 = \alpha^2 + \alpha + 2\alpha c + c
\end{aligned}$$

and also we get,

$$\hat{H}''_{\alpha}(\lambda) = \frac{\alpha(\alpha+1)}{(\lambda+1)^{\alpha+2}} \hat{h}(\lambda+1) - \frac{2\alpha}{(\lambda+1)^{\alpha+1}} \hat{h}'(\lambda+1) + \frac{1}{(\lambda+1)^{\alpha}} \hat{h}''(\lambda+1)$$

$$\begin{aligned}
\Rightarrow |\lambda^2 \hat{H}''_{\alpha}(\lambda) &\leq |\frac{\hat{h}(\lambda+1)}{(\lambda+1)^{\alpha}}|\{|\frac{\lambda^2\alpha(\alpha+1)}{(\lambda+1)^2}| + c|\frac{2\alpha\lambda^2}{(\lambda+1)^2}| + c|\frac{\lambda^2}{(\lambda+1)^2}|\} \\
&\leq |\frac{\hat{h}(\lambda+1)}{(\lambda+1)^{\alpha}}|(\alpha^2 + \alpha + 2\alpha c + c) \\
&\leq c_1 |\hat{H}_{\alpha}(\lambda)|
\end{aligned}$$

thus $H_{\alpha}(t)$ is 2-regular. Now for the sectorial of $H_{\alpha}(t)$ we have

$$|arg(\hat{H}_{\alpha}(\lambda))| = |arg \frac{1}{(\lambda+1)^{\alpha}} \hat{h}(\lambda+1)| \leq \theta < \frac{\pi}{2} \qquad \forall Re(\lambda) > 0.$$

Thus $H_\alpha(t)$ is θ-sectorial with $\theta < \frac{\pi}{2}$.
Then (11) is parabolic equation and from Theorem 2.1 and Proposition 2.1 we deduce that equation (11) admits a unique bounded resolvent $S_\alpha \in C^1(J; B(X))$ satisfies for some $M \geq 1$ the two equations (8) and (9). And the solution of (11) will be given in the form $v_\alpha(t) = \frac{d}{dt}(S_\alpha(t)) * F_\alpha(t))$ and where $f \in W^{1,1}(J; X_A),\ u_o \in X_A$, we get

$$F_\alpha(t) = e^{-t}u_o + e^{-t}\phi_\alpha(t) * e^{-t}f(t) \Rightarrow F_\alpha \in W^{1,1}(J; X_A)$$

which gives by proposition 2.2 that $v_\alpha(t)$ given by

$$v_\alpha(t) = S_\alpha(t)F_\alpha(0) + \int_0^t S_\alpha(t-s)\dot{F}_\alpha(s)ds, \qquad t \in J \tag{12}$$

is a strong solution of (11).
Now we have

$$\dot{F}_\alpha(t) = -e^{-t}u_o + e^{-t}\phi_\alpha(t)f(0) + e^{-t}\phi_\alpha(t) * (e^{-t}\dot{f}(t) - e^{-t}f(t))$$

and we can write (12) in the form

$$\begin{aligned} v_\alpha(t) &= S_\alpha(t)u_o - S_\alpha(t) * e^{-t}u_o + S_\alpha(t) * e^{-t}\phi_\alpha(t)f(0) + S_\alpha(t) * e^{-t}\phi_\alpha(t) * e^{-t}\dot{f}(t) \\ &\quad -S_\alpha(t) * e^{-t}\phi_\alpha(t) * e^{-t}f(t). \end{aligned}$$

But we have $v_\alpha(t) = e^{-t}u_\alpha(t)$ thus we can get $u_\alpha(t)$ in the form

$$u_\alpha(t) = e^tS_\alpha(t)u_o - e^tS_\alpha(t) * u_o + e^tS_\alpha(t) * \phi_\alpha(t) * (f(0)\delta(t) + \dot{f}(t) - f(t)). \tag{13}$$

Now we want to prove that (13) satisfies (2).

$$\begin{aligned} Du_\alpha(t) &= e^tS_\alpha(t)u_o + e^t\dot{S}_\alpha(t)u_o - e^tS_\alpha(t)u_o \\ &\quad +\phi_\alpha(t) * (f(0)\delta(t) + \dot{f}(t) - f(t)) \\ &\quad +(e^tS_\alpha(t) + e^t\dot{S}_\alpha(t)) * \phi_\alpha(t) * (f(0)\delta(t) + \dot{f}(t) - f(t)) \end{aligned}$$

Now we can write $D^\alpha u_\alpha(t) = \phi_{1-\alpha}(t) * Du_\alpha(t)$ which gives

$$\begin{aligned} D^\alpha u_\alpha(t) &= \phi_{1-\alpha}(t) * e^t\dot{S}_\alpha(t)u_o + \phi_1(t) * (f(0)\delta(t) + \dot{f}(t) - f(t)) \\ &\quad +\phi_1(t) * D(e^tS_\alpha(t)) * (f(0)\delta(t) + \dot{f}(t) - f(t)) \\ &= \phi_{1-\alpha}(t) * e^t\dot{S}_\alpha(t)u_o + e^tS_\alpha(t) * (f(0)\delta(t) + \dot{f}(t) - f(t)) \end{aligned}$$

But we have

$$\begin{aligned} e^tS_\alpha(t) &= e^t + h_\alpha(t) * Ae^tS_\alpha(t), \\ e^t\dot{S}_\alpha(t)x &= h_\alpha(t)Ax + Ah_\alpha(t) * e^t\dot{S}_\alpha(t)x, \quad x \in D(A) \end{aligned}$$

Then we get the following

$$\begin{aligned} D^\alpha u_\alpha(t) &= \phi_{1-\alpha}(t) * [h_\alpha(t)Au_o + Ah_\alpha(t) * e^t\dot{S}_\alpha(t)u_o] \\ &\quad +(e^t + h_\alpha(t) * Ae^tS_\alpha(t)) * (f(0)\delta(t) + \dot{f}(t) - f(t)) \end{aligned}$$

$$\begin{aligned}
&= \phi_1(t) * h(t)Au_o + A\phi_1(t) * h(t) * e^t\dot{S}_\alpha(t)u_o \\
&\quad + e^t * (f(0)\delta(t) + \dot{f}(t) - f(t)) \\
&\quad + A(\phi_\alpha(t) * h(t) * e^t S_\alpha(t) * (f(0)\delta(t) + \dot{f}(t) - f(t)) \\
&= h(t) * Au_o + h(t) * A\phi_1(t) * (e^t\dot{S}_\alpha(t)u_o + e^t S_\alpha(t)u_o) \\
&\quad - h(t) * \phi_1(t) * Ae^t S_\alpha(t)u_0 + e^t * (f(0)\delta(t) + \dot{f}(t) - f(t)) \\
&\quad + \phi_\alpha(t) * h(t) * Ae^t S_\alpha(t) * (f(0)\delta(t) + \dot{f}(t) - f(t)).
\end{aligned}$$

But we have $e^t * (f(0)\delta(t) + \dot{f}(t) - f(t)) = f(t)$ then we get

$$\begin{aligned}
D^\alpha u_\alpha(t) &= f(t) + h(t) * A\{e^t S_\alpha(t)u_o - e^t S_\alpha(t) * u_o \\
&\quad + e^t S_\alpha(t) * \phi_\alpha(t) * (f(0)\delta(t) + \dot{f}(t) - f(t))\}
\end{aligned}$$

and from (13) we get $u_\alpha(0) = u_o$ thus $u_\alpha(t)$ given by (13) is the unique strong solution of problem (2).

For the mild solution we have the following corollary

Corollary 3.1 Replacing $f \in W^{1,1}(J, X_A)$ in Theorem 3.1 by $f \in W^{1,1}(J, X)$. Then there is a unique mild solution $u_\alpha \in C(J, X)$ of (2) given by (13).

Proof. Where $f(t) \in W^{1,1}(J, X)$, we get

$$F_\alpha(t) = e^{-t}u_o + e^{-t}\phi_\alpha(t) * e^{-t}f(t) \in W^{1,1}(J, X)$$

then from Proposition 2.2 we get that (12) is a mild solution of (11) which gives $H_\alpha * v_\alpha \in C(J, X_A)$ with $v_\alpha(t) = F_\alpha(t) + A(H_\alpha * v_\alpha)(t)$ and where $H_\alpha(t) = e^{-t}(\phi_\alpha(t) * h(t)$, $F_\alpha(t) = e^{-t}u_o + e^{-t}\phi_\alpha(t) * e^{-t}f(t)$ and $v_\alpha(t) = e^{-t}u_\alpha(t)$ we get, $h_\alpha * u_\alpha \in C(J; X_A)$ and $u_\alpha(t) = u_o + A(h_\alpha * u_\alpha)(t) + (\phi_\alpha * f)(t)$ on J, thus (13) is a mild solution of (2).

4. Continuation Theorem

Let $u_1(t)$ be the solution of the problem

$$\frac{du(t)}{dt} = f(t) + \int_0^t h(t-s)Au(s)ds, \quad u(0) = u_o. \tag{14}$$

As the proof of Theorem 3.1 and Corollary 3.1 we can prove the following theorem
Theorem 4.1 Let $u_o \in D(A), f \in W^{1,1}(J, X_A)$ with $h(t)$ is 2-regular and A is the infinitesimal generator of a bounded C_o semigroup $\{T(t),\ t \geq 0\}$.If $|arg(\frac{\hat{h}(\lambda+1)}{(\lambda+1)}| \leq \theta < \frac{\pi}{2}$, for $Re\lambda \geq 0$ then there is a unique strong solution $u_1(t) \in C(J, X_A)$ of (14) given by

$$u_1(t) = e^t S_1(t)u_o - e^t S_1(t) * u_o + e^t S_1(t) * \phi_1(t) * (f(0)\delta(t) + \dot{f}(t) - f(t)) \tag{15}$$

where $S_1(t) \in C^1(J; B(X))$ is a bounded resolvent given as in Definition 2.4, with $a(t) = H_1(t) = e^{-t}h_1(t) = e^{-t}(\phi_1(t) * h(t))$, and satisfies Theorem 2.1.

If we put $f(t) \in W^{1,1}(J,X)$ instead of $f(t) \in W^{1,1}(J,X_A)$ then there is a unique mild solution of (14) given by (15).

Theorem 4.2 If the solution of the initial value problem (2) exists, then

$$\lim_{\alpha \to 1^-} u_\alpha(t) = u_1(t) \tag{16}$$

Proof. Taking Laplace transformation to both sides of (15) and (13), and using the inequality

$$(e^t S_\alpha(t))\hat{} = (I - \frac{\hat{h}(\lambda)}{\lambda^\alpha} A)^{-1}/(\lambda - 1) \text{ and } (e^t S_1(t))\hat{} = (I - \frac{\hat{h}(\lambda)}{\lambda} A)^{-1}/(\lambda - 1), \quad Re(\lambda) > 1$$

where ($\hat{}$) denotes the Laplace transformation, we get

$$\hat{u}_\alpha(\lambda) = \frac{1}{\lambda}(I - \frac{\hat{h}(\lambda)}{\lambda^\alpha} A)^{-1} u_o + \frac{1}{\lambda^\alpha}(I - \frac{\hat{h}(\lambda)}{\lambda^\alpha} A)^{-1} \hat{f}(\lambda) \tag{17}$$

$$\hat{u}_1(\lambda) = \frac{1}{\lambda}(I - \frac{\hat{h}(\lambda)}{\lambda} A)^{-1} u_o + \frac{1}{\lambda}(I - \frac{\hat{h}(\lambda)}{\lambda} A)^{-1} \hat{f}(\lambda) \tag{18}$$

eliminate $\hat{f}(\lambda)$ from (17) and (18) we get,

$$(\lambda^\alpha I - \hat{h}(\lambda) A)\hat{u}_\alpha(\lambda) - \frac{1}{\lambda^{1-\alpha}} u_o = (\lambda I - \hat{h}(\lambda) A)\hat{u}_1(\lambda) - u_o$$

$$\Rightarrow (\frac{1}{\lambda^{1-\alpha}} - 1)\hat{u}_\alpha(\lambda) + (I - \frac{\hat{h}(\lambda)}{\lambda} A)\hat{u}_\alpha(\lambda) - \frac{1}{\lambda^{2-\alpha}} u_o = (I - \frac{\hat{h}(\lambda)}{\lambda} A)\hat{u}_1(\lambda) - \frac{1}{\lambda} u_o.$$

Operating by $(I - \frac{\hat{h}(\lambda)}{\lambda} A)^{-1}$ we get

$$\hat{u}_1(\lambda) - \hat{u}_\alpha(\lambda) = (I - \frac{\hat{h}(\lambda)}{\lambda} A)^{-1}(\frac{1}{\lambda^{1-\alpha}} - 1)\hat{u}_\alpha(\lambda) + (I - \frac{\hat{h}(\lambda)}{\lambda} A)^{-1}(\frac{1}{\lambda} - \frac{1}{\lambda^{2-\alpha}}) u_o.$$

$$\begin{aligned}
\hat{u}_1(\lambda) - \hat{u}_\alpha(\lambda) &= [(I - \frac{\hat{h}(\lambda)}{\lambda} A)^{-1}/(\lambda - 1)] \times \\
&\quad [(\lambda - 1)(\frac{1}{\lambda^{1-\alpha}} - 1)\hat{u}_\alpha(\lambda) + (\lambda - 1)(\frac{1}{\lambda} - \frac{1}{\lambda^{2-\alpha}}) u_o] \\
&= (e^t S_1(t))\hat{}[(\frac{1}{\lambda^{1-\alpha}} - 1)(\lambda \hat{u}_\alpha(\lambda) - u_o) \\
&\quad -(\frac{1}{\lambda^{1-\alpha}} - 1)\hat{u}_\alpha(\lambda) - (\frac{1}{\lambda} - \frac{1}{\lambda^{2-\alpha}}) u_o]
\end{aligned}$$

By taking Laplace inverse transformation we get

$$u_1(t) - u_\alpha(t) = e^t S_1(t) * [(\phi_{1-\alpha}(t) - \delta(t)) * (Du_\alpha(t) - u_\alpha(t)) - (\phi_1(t) - \phi_{2-\alpha}(t)) u_o]$$

if we let

$$R_\alpha(t) = [(\phi_{1-\alpha}(t) - \delta(t)) * (Du_\alpha(t) - u_\alpha(t)) - (\phi_1(t) - \phi_{2-\alpha}(t)) u_o] \tag{19}$$

then by taking the norm $\|.\|$ using $\|S_1(t)\| \leq M$ for all $t > 0$ we get

$$\|u_1(t) - u_\alpha(t)\| \leq Me^T \int_0^t \|R_\alpha(s)\| ds$$

but we have from (19) that

$$\|R_\alpha(t)\| \to 0 \quad \text{as} \quad \alpha \to 1^-$$

which gives that

$$\|u_1(t) - u_\alpha(t)\| \to 0 \quad \text{as} \quad \alpha \to 1^-$$

Theorem 4.3 If the solution of the initial value problem (6) exists with $u_o \in D(A^2)$ and $f \in W^{1,1}(J, D(A^2))$ we get

$$\lim_{\alpha \to 1^-} D^\alpha u_\alpha(t) = \frac{du_1(t)}{dt} \tag{20}$$

Proof. We have from (13) that

$$Au_\alpha(t) = A\left[e^t S_\alpha(t)u_o - e^t S_\alpha(t) * u_o + e^t S_\alpha(t) * \phi_\alpha(t) * (f(0)\delta(t) + \dot{f}(t) - f(t))\right].$$

Let $Au_o = v_o \in X_A, \quad Af = g \in W^{1,1}(J; X_A)$ then we get

$$Au_\alpha(t) = e^t S_\alpha(t)v_o - e^t S_\alpha(t) * v_o + e^t S_\alpha(t) * \phi_\alpha(t) * (g(0)\delta(t) + \dot{g}(t) - g(t)). \tag{21}$$

But by Theorem 3.2 we have that (21) is a strong solution of

$$D^\alpha v(t) = g(t) + \int_0^t h(t-s)Av(s)ds, \quad v(0) = v_o.$$

We prove in Theorem 4.2 that this solution converges to $v_1(t) \in X_A$, A is a closed operator with $u_\alpha(t) \to u_1(t)$ and $Au_\alpha(t) \to v_1(t)$ then $v_1(t) = Au_1(t)$ thus we get $Au_\alpha(t) \to Au_1(t)$ as $\alpha \to 1^-$. Now we have

$$\|D^\alpha u_\alpha(t) \quad Du_1(t)\| = \|h(t) * (Au_\alpha(t) \quad Au_1(t))\|$$

then

$$\|D^\alpha u_\alpha(t) - Du_1(t)\| \leq |h(t)| \|Au_\alpha(t) - Au_1(t)\| \to 0 \text{ as } \alpha \to 1^-$$

thus we get $\lim_{\alpha \to 1^-} D^\alpha u_\alpha(t) = \dfrac{du_1(t)}{dt}$

5. Maximal Regularity

Consider now the Cauchy problem

$$D^\alpha u(t) = f(t) + \int_0^t h(t-s)Au(s)ds, \qquad u(0) = 0, \quad t \in [0, 1] \tag{22}$$

which gives the strong solution given by (13) in the form

$$u_\alpha(t) = e^t S_\alpha(t) * \phi_\alpha(t) * (f(0)\delta(t) + \dot{f}(t) - f(t)).$$

The maximal regularity result that we prove states essentially that if f is $\delta-$Hölder continuous then there is a unique solution $u_\alpha(t)$ such that $D^\alpha u_\alpha$ and Au_α are $\delta-$Hölder continuous (see [10,16]).

Theorem 5.1 For problem (22) if $f \in C_0^\delta([0,1];X)$ for $\delta \in (0,1)$ with the condition $(f*\phi_\alpha)(t) \in X_A$ for all $t \in J=[0,1]$, then there is a strong solution $u_\alpha \in C_0^\delta([0,1];X_A)$ where $D^\alpha u_\alpha, Au_\alpha \in C_0^\delta([0,1];X)$.

Proof. Since $F_\alpha(t) = e^{-t}f_\alpha(t) = e^{-t}(\phi_\alpha(t)*f(t))$ and $(f*\phi_\alpha)(t) \in X_A$ $\Rightarrow F_\alpha(t) \in X_A$ for all $t \in [0,1]$. Now we have

$$\begin{aligned}
\|f_\alpha(t+\tau)-f_\alpha(t)\| &= \|\int_0^{t+\tau}\phi_\alpha(t+\tau-s)f(s)ds - \int_0^t \phi_\alpha(t-s)f(s)ds\| \\
&= \|\int_0^{\tau}\phi_\alpha(t+\tau-s)f(s)ds + \int_0^t \phi_\alpha(t-s)(f(s+\tau)-f(s))ds\| \\
&\le \int_0^t M\frac{(t+\tau-s)^{\alpha-1}}{\Gamma(\alpha)}ds + \int_0^t \frac{(t-s)^{\alpha-1}}{\Gamma(\alpha)}\|f(s+\tau)-f(s)\|ds \\
&\le \frac{M}{\Gamma(\alpha+1)}((t+\tau)^\alpha - t^\alpha) + \int_0^t \frac{(t-s)^{\alpha-1}}{\Gamma(\alpha)}\|f(s+\tau)-f(s)\|ds
\end{aligned}$$

But we have from the mean value theorem that

$$\|(t+\tau)^\alpha - t^\alpha\| = \alpha c^{\alpha-1}\tau \le \alpha c^{\alpha-1}\tau^\delta, \quad 0<t<c<t+\tau<1$$

and where $f \in C_0^\delta([0,1];X)$ we get $f_\alpha(t) = (\phi_\alpha * f)(t) \in C_0^\delta([0,1];X_A)$. But since $F_\alpha(t) = e^{-t}f_\alpha(t)$ then we get

$$\|F_\alpha(t)-F_\alpha(s)\| \le |e^{-t}-e^{-s}|\|f_\alpha(t)\| + e^{-s}\|f_\alpha(t)-f_\alpha(s)\|$$

but by the inequality $|e^{\lambda t}-e^{\lambda s}| \le (1-\delta)^{1-\delta}\frac{|\lambda|}{|Re\lambda|^{1-\delta}}|t-s|^\delta, \quad Re\lambda<0$ we find that $F_\alpha(t) \in C_0^\delta(J;X)$. Now from Theorem 2.2 we get that there is a strong solution $v_\alpha \in C_0^\delta([0,1],X_A) \Rightarrow u_\alpha \in C_0^\delta([0,1],X_A)$ where $u_\alpha(t) = e^t v_\alpha(t)$ and for the fractional derivative we get

$$\begin{aligned}
\|D^\alpha u_\alpha(t+\tau) - D^\alpha u_\alpha(t)\| &= \|f(t+\tau)-f(t) + \int_0^t h(s)A(u_\alpha(s+\tau)-u_\alpha(s))ds\| \\
&\le \|f(t+\tau)-f(t)\| + \int_0^t |h(s)|\|A(u_\alpha(s+\tau)-u_\alpha(s))\|ds \\
&\le M\tau^\delta + \int_0^t |h(s)|\|u_\alpha(s+\tau)-u_\alpha(s)\|_A ds
\end{aligned}$$

and where $u_\alpha \in C_0^\delta(J;X_A)$, $h \in L^1_{loc}(R^+)$ we get that $D^\alpha u_\alpha \in C_0^\delta([0,1],X)$ and where $u_\alpha \in C_0^\delta([0,1];X_A)$ it is obvious that $Au_\alpha(t) \in C_0^\delta([0,1],X)$.

6. Examples

Here we give two examples of our Problem (2) for the closed operator A and give the solution to two special cases for $h(t)$ with $f(t)=0$.

Example 1 :

Let the operator A be defined by

$$D(A) = \{u(x,t) \in C^1(0,\infty) \mid u(0,t) = 0\}, \quad Au(x,t) = \frac{\partial}{\partial x}u(x,t),$$

then (2) takes the form

$$D^\alpha u(x,t) = h(t) * \frac{\partial}{\partial x}u(x,t) + f(x,t), \qquad u(0,x) = u_o(x), \;\; u(t,0) = 0. \tag{23}$$

Now we know that the operator A generates a contraction semigroup (see[17]) which shows that if $h(t)$ is 2-regular satisfying that $|arg(\frac{\hat{h}(\lambda+1)}{(\lambda+1)^\alpha})| \leq \theta < \frac{\pi}{2}$ then we can apply Theorem 3.1 and then (23) has a unique strong solution if $f \in W^{1,1}(J; X_A)$ and a unique mild solution if $f \in W^{1,1}(J; X)$ given by (13). Now with $f(x,t) = 0$ we can write (23) as follows

$$\phi_{1-\alpha}(t) * Du(x,t) = h(t) * \frac{\partial}{\partial x}u(x,t), \tag{24}$$

taking Laplace transformation for the variable t with parameter λ, we get

$$\frac{1}{\lambda^{1-\alpha}}(\lambda u^*(x,\lambda) - u(x,0)) = h^*(\lambda)\frac{\partial}{\partial x}u^*(x,\lambda), \quad u^*(0,\lambda) = 0$$

$$\Rightarrow \lambda^\alpha u^*(x,\lambda) - \lambda^{\alpha-1}u_o(x) = h^*(\lambda)\frac{\partial}{\partial x}u^*(x,\lambda), \quad u^*(0,\lambda) = 0$$

where $u^*(x,\lambda)$ is the Laplace transformation of $u(x,t)$ for the variable t.
Now taking Laplace transformation for the variable x with parameter ν we get

$$\lambda^\alpha U^*(\nu,\lambda) - \lambda^{\alpha-1}U_o(\nu) = h^*(\lambda)(\nu U^*(\nu,\lambda) - u^*(0,\lambda)),$$

where $U^*(\nu,\lambda)$ is the Laplace transformation of $u^*(x,\lambda)$ for the variable x.
Which implies that,

$$U^*(\nu,\lambda) = \frac{\lambda^{\alpha-1}}{\lambda^\alpha - \nu h^*(\lambda)}U_o(\nu). \tag{25}$$

Now we consider the following special cases.

Case 1. Take $h(t) = \phi_{1-\alpha}(t)$ we can easily prove that h(t) is 2-regular and it satisfies that $|arg(\frac{\hat{h}(\lambda+1)}{(\lambda+1)^\alpha})| = |arg(\frac{1}{\lambda+1})| \leq \theta < \frac{\pi}{2}$ for $Re(\lambda) > 0$ thus it satisfies Theorem 3.1. Then (25) takes the form

$$U^*(\nu,\lambda) = \frac{\lambda^{\alpha-1}}{\lambda^\alpha - \frac{\nu}{\lambda^{1-\alpha}}} = \frac{1}{\lambda-\nu}U_o(\nu).$$

Taking the inverse Laplace transformation for λ, we get

$$U(\nu,t) = e^{\nu t}U_o(\nu),$$

then taking Laplace transform for ν gives the solution in the form

$$u(x,t) = \frac{1}{2\pi i}\int_{\mu-i\infty}^{\mu+i\infty} e^{\nu t}e^{\nu x}U_o(\nu)d\nu = \frac{1}{2\pi i}\int_{\mu-i\infty}^{\mu+i\infty} e^{\nu(t+x)}U_o(\nu)d\nu = u_o(x+t)$$

which is the know solution to the problem

$$\frac{\partial u(x,t)}{\partial t} = \frac{\partial u(x,t)}{\partial x}, \quad t,x>0, \;\; u(0,t)=0, \;\; u(x,0)=u_o(x).$$

Case 2. Take $h(t) = \phi_{\gamma-\alpha}(t),\ \gamma-\alpha \in (0,1),\ \gamma,\alpha \in (0,1)$, we can easily prove that h(t) is 2-regular and it satisfies that $|arg(\frac{\hat{h}(\lambda+1)}{(\lambda+1)^\alpha})| = |arg(\frac{1}{(\lambda+1)^\gamma})| \leq \theta < \frac{\pi}{2}$ for $Re(\lambda) > 0$ thus it satisfies Theorem 3.1. Then (25) takes the form

$$U^*(\nu,\lambda) = \frac{\lambda^{\alpha-1}}{\lambda^\alpha - \frac{\nu}{\lambda^{\gamma-\alpha}}} = \frac{\lambda^{\gamma-1}}{\lambda^\gamma-\nu}U_o(\nu),$$

taking the inverse Laplace transform for λ gives

$$U(\nu,t) = E_{\gamma,1}(\nu t^\gamma)U_o,$$

where $E_{\gamma,1}(\nu t^\gamma)$ is the Mittag-Leffler function (see[13])
then take the inverse Laplace transform for ν, we get the solution in the form

$$u(x,t) = G(x,t) * u_o(x)$$

with

$$G(x,t) = L^{-1}(E_{\gamma,1}(\nu t^\gamma)) = \frac{1}{2\pi i}\int_{\mu-i\infty}^{\mu+i\infty} e^{\nu t}E_{\gamma,1}(\nu t^\gamma)d\nu.$$

Example 2 :

Let the operator A be defined by

$$D(A) = \{u(x,t) \in C^2(-\infty,\infty),\ \lim_{x\to\pm\infty} u(x,t) = 0\},\ Au(x,t) = \frac{\partial^2}{\partial x^2}u(x,t)$$

then our problem takes the form

$$D^{\alpha}u(x,t) = h(t)*\frac{\partial^2}{\partial x^2}u(x,t)+f(x,t),\ |x|<\infty,\ t>0,\qquad u(x,0)=u_o(x),\ \lim_{x\to\pm\infty}u(x,t)=0 \tag{26}$$

Now we know that the operator A generates a contraction semigroup (see[17]) which shows that if $h(t)$ is 2-regular satisfying that $|arg(\frac{\hat{h}(\lambda+1)}{(\lambda+1)^{\alpha}})| \leq \theta < \frac{\pi}{2}$ then we can apply Theorem 3.1 and then (26) has a unique strong solution if $f \in W^{1,1}(J;X_A)$ and a unique mild solution if $f \in W^{1,1}(J;X)$ given by (13). Now with $f(x,t) = 0$ we can write (26) as follows

$$D^{\alpha}u(x,t) = h(t)*\frac{\partial^2}{\partial x^2}u(x,t),\ |x|<\infty,\ t>0,\qquad u(x,0)=u_o(x),\ \lim_{x\to\pm\infty}u(x,t)=0. \tag{27}$$

Taking Laplace transform for the variable t with parameter λ for (27) we get

$$\lambda^{\alpha}u^*(x,\lambda)-\lambda^{\alpha-1}u_o(x) = h^*(\lambda)\frac{\partial^2}{\partial x^2}u^*(x,\lambda),\qquad \lim_{x\to\pm\infty}u(x,\lambda)=0,$$

where $u^*(x,\lambda)$ is the Laplace transformation of $u(x,t)$ for the variable t.
Now taking Fourier transform for the variable x with parameter ν, we get,

$$\lambda^{\alpha}U^*(\nu,\lambda)-\lambda^{\alpha-1}U_o(\nu) = -h^*(\lambda)\nu^2U^*(\nu,\lambda)$$

where $U^*(\nu,\lambda)$ is the Fourier transformation of $u^*(x,\lambda)$ for the variable x.
Which implies that,

$$U^*(\nu,\lambda) = \frac{\lambda^{\alpha-1}}{\lambda^{\alpha}+\nu^2h^*(\lambda)}U_o(\nu). \tag{28}$$

Now we consider the two special cases given in Example 1.

Case 1. Take $h(t) = \phi_{1-\alpha}(t)$ in (28) we get

$$U^*(\nu, \lambda) = \frac{1}{\lambda + \nu^2} U_o(\nu),$$

take the inverse Laplace transform, we get

$$U(\nu, t) = e^{-\nu^2 t} U_o(\nu),$$

then take the inverse Fourier transform, we get

$$u(x,t) = G(x,t) * u_o(x)$$

where

$$G(x,t) = \frac{1}{2\pi} \int_{-\infty}^{\infty} e^{-\nu^2 t} e^{-i\nu x} d\nu = \frac{1}{2\pi} e^{\frac{-x^2}{4t}} \int_{-\infty}^{\infty} e^{-(\sqrt{t}\nu - \frac{ix}{2\sqrt{t}})^2} d\nu$$

putting $\sqrt{t}\nu - \frac{ix}{2\sqrt{t}} = z$ we obtain

$$G(x,t) = \frac{1}{2\pi\sqrt{t}} e^{\frac{-x^2}{4t}} \int_{-\infty}^{\infty} e^{-z^2} dz = \frac{1}{2\sqrt{\pi t}} e^{\frac{-x^2}{4t}}.$$

Thus the solution is given by

$$u(x,t) = \frac{1}{2\sqrt{\pi t}} e^{\frac{-x^2}{4t}} * u_o(x)$$

which is the known solution to the problem

$$\frac{\partial u}{\partial t} = \frac{\partial^2 u}{\partial x^2} \quad with \quad u(x,0) = u_o(x), \quad \lim_{x \to \pm\infty} u(x,t) = 0.$$

Case 2. Take $h(t) = \phi_{\gamma-\alpha}(t)$ in (28) we get

$$U^*(\nu, \lambda) = \frac{\lambda^{\gamma-1}}{\lambda^\gamma + \nu^2} U_o(\nu),$$

take the inverse Laplace transform, we get

$$U(\nu, t) = E_{\gamma,1}(-\nu^2 t^\gamma) U_o(\nu),$$

then take the inverse Fourier transform, we get

$$u(x,t) = G(x,t) * u_o(x)$$

where

$$G(x,t) = \frac{1}{2\pi} \int_{-\infty}^{\infty} e^{i\nu x} E_{\gamma,1}(-\nu^2 t^\gamma) d\nu.$$

References

[1] M. Caputo, Linear model of dissipation whose Q is almost frequency independent-II, *Geophys. J. R. Astr. Soc.* Vol. 13, pp. 529-539. (1967)

[2] A. M. A. El-Sayed, Fractional order evolution equations. *J. of Frac. Calculus* Vol. 7, pp. 89-100. (1995).

[3] A. M. A. El-Sayed, Fractional-order diffusion-wave equation. *Int. J. Theoretical Physics.* Vol. 35 (2) , pp. 311-322. (1996)

[4] A. M. A. El-Sayed, Fractional-order evolutionary integral equations, *Appl. Math. Comput.* Vol. 98, No. 2-3, pp. 139-146. (1999)

[5] A. M. A. El-Sayed and Mohamed. A. E. Aly, On the continuation of fractional order evolutionary integral equations and some applications, *The Korean Journal of Computational and Applied Mathematics* Vol.9, No.2, pp 525-533, 2002

[6] A. M. A. El-Sayed and Mohamed. A. E. Herzallah,Continuation and maximal regularity of fractional-order evolution equation, *J. Math. Anal. Appl.*, Vol.296, Issue 1, pp 340-350,(2004).

[7] I.M. Gelfand and G.E. Shilov. *Generalized functions,* Vol. 1, Moscow (1958).

[8] R. Gorenflo and F. Mainardi, *Fractional calculus: integral and differential equations of fractional order*. In: Fractals and Fractional Calculus in Continuum Mechanics (A. Carpinteri and F. Mainardi, Ed-s), Springer Verlag - Wien - New York (1997), 223-276.

[9] R.C. Grimmer and A.J. Prichard, *Analytic resolvent operators for integral equations in Banach spaces*, *J. Diff. Equat.*, Vol. 50 (1983), 234-259.

[10] A. Lunardi, *Analytic Semigroups and Optimal Regularity in Parabolic Problems,* Birkhäuser, Basel, 1995.

[11] K. S. Miller and B. Ross, *An Introduction to the Fractional Calculus and Fractional Differential Equations,* John Wiley & Sons. Inc., New York 1993.

[12] I. Podlubny and A. M.A. El-Sayed, *On Two Definitions Of Fractional Calculus*, Slovak Academy Of Sciences Institute Of Experimental Physics U E F - 03 -96 ISBN 80-7099- 252-2. (1996).

[13] I. Podlubny, *Fractional Differential Equation.* Acad. Press, San Diego - New York - London, 1999.

[14] J. Prüss, *Evolutionary Integral Equations and Applications.* Birkhauser Verlag Basel, Boston, Berlin 1993.

[15] S.G. Samko, A.A. Kilbas and O.I. Marichev. *Integral and derivatives of the fractional orders and some of their applications.* Nauka i Tekhnika Minisk, (1987).

[16] E. Sinestrari, On the abstract Cauchy problem in spaces of continuous functions, *J. Math. Anal. Appl.* Vol. 107, 16-66, (1985).

[17] K. Yosida *Functional Analysis,* 3rd ed. Springer Verlag, 1971.

INDEX

N

O

P

Q

R

S

T

U

V

Y